Statistical Tables

THIRD EDITION

Statistical Tables

THIRD EDITION

F. James Rohlf

Robert R. Sokal

State University of New York
at Stony Brook

W. H. Freeman and Company
New York

Library of Congress Cataloging-in-Publication Data

Rohlf, F. James, 1936–
    Statistical tables / F. James Rohlf, Robert R. Sokal.—3d ed.
      p. cm.
    Collection of tables to accompany Biometry, 3d ed.
    ISBN 0-7167-2412-X
    1. Biometry—Tables.   2. Mathematical statistics—Tables.   I. Sokal, Robert R.
II. Sokal, Robert R. Biometry. III. Title.
QH323.5.R63 1995
519.5′0212—dc20   94-11121
                                                            CIP

Printed in the United States of America

Second printing 1995, VB

To Kathryn, Karen, David, and Hannah

Contents

Preface

This set of tables grew out of our dissatisfaction with the customary placement of statistical tables at the end of textbooks of biometry and statistics. Serious users of these books and tables are constantly inconvenienced by having to turn back and forth between the text material on a certain method and the table necessary for the test of significance or for another computational step. Occasionally the tables are interspersed throughout a textbook at sites of their initial application; they are then difficult to locate, and users still must turn back and forth in the book. Frequent users of statistics, therefore, generally use one or more sets of statistical tables, not only because these usually contain more complete and diverse statistical tables than do the textbooks, but also because they eliminate the need to constantly turn pages from text to tables.

When we first planned to write our biometry textbook (Sokal and Rohlf, *Biometry*, Third Edition, W. H. Freeman and Company, New York), we thought to eliminate tables altogether, asking readers to furnish their own statistical tables from those available. For pedagogical reasons, however, we found it desirable to refer to a standard set of tables, so we undertook to furnish such tables separately from the text. We gave considerable thought to making these tables as useful as we could for statistical work in the biological and social sciences.

Several of the tables would have been very complicated and tedious to recompute. These have been copied with permission of authors and publishers whose courtesy is here acknowledged collectively. We are indebted to the Literary Executor of the late Sir Ronald A. Fisher, F.R.S., Cambridge, to Dr. Frank Yates, F.R.S., Rothamsted, and to Messrs. Oliver and Boyd, Limited, Edinburgh, for permission to reprint tables III and XX from their book *Statistical Tables for Biological, Agricultural and Medical Research*. Other specific acknowledgments are found beneath the particular table. We appreciate the constructive comments of Professor K. R. Gabriel, University of Rochester, who read a draft of the introductory material. We appreciate the dedicated efforts of Barbara Thomson, who helped us put together the revised materials for this edition.

We hope users of statistics will find these tables as useful as we have already found them to be in our work. We will be grateful for any suggestions about changes, additions, or deletions, as well as for any corrections.

F. James Rohlf
Robert R. Sokal
July 1994

Notes on the Third Edition

This edition of *Statistical Tables* is substantially different from its predecessors. In an age in which many mathematical functions are readily available on pocket calculators or personal computers, it no longer makes sense to furnish them in statistical tables. We have therefore removed eight tables from the previous edition, all of which can easily be replicated with a modern pocket calculator. We have replaced them with seven new tables, including two tables for critical values of the δ-corrected one-sample Kolmogorov–Smirnov test; shortest unbiased confidence limits for the Poisson and binomial distributions; critical values for several selected proportions (including 0.5, which is useful for the sign test); nomographs for power and sample size in analysis of variance; and a table for a test of serial independence of observations.

An introduction on interpolation precedes the tables. Each table is accompanied by a brief explanation, a demonstration of how to look up a value, and a reference to the section or sections in our textbook, *Biometry,* Third Edition, giving explanations and applications of the table. All references to section, table, or box numbers unaccompanied by a citation pertain to the textbook. (Those who use the set of tables but not the textbook should simply disregard these references.) Each table is accompanied by a short explanation of how it was generated. Most of the tables were generated using simple computer programs. In such cases equations are given to explain how the tables were prepared.

F. James Rohlf
Robert R. Sokal
July 1994

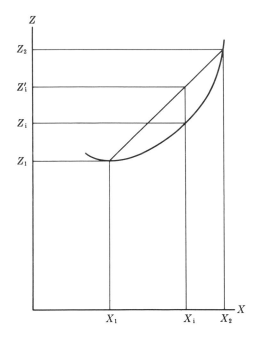

Introduction: Interpolation

Finding a value of a function for an argument that is intermediate between two arguments in a table requires *interpolation*. In some tables published earlier, aids to mental interpolation (proportional parts) are furnished. Since the present tables are oriented toward use with calculators, we furnish several formulas especially adapted for their use.

We shall employ the following symbolism. The tabled arguments to each side of the desired argument X_i are identified as X_1 and X_2, respectively. Argument X_i must lie between the tabled arguments, $X_1 < X_i < X_2$ or $X_1 > X_i > X_2$. The functions shown in the table are Z_1 and Z_2 corresponding to X_1 and X_2 and the desired function corresponding to argument X_i is labeled Z_i.

The simplest method is *linear interpolation*. It assumes that the function $Z = f(X)$ is approximately linear over the interval from X_1 to X_2. It serves as an adequate method where the interval over which one needs to interpolate is not very wide, or where the function is either truly or approximately linear in that interval. The effect of linear interpolation is seen in the accompanying figure, which illustrates a linear function approximating a curvilinear function over the interval from X_1 to X_2. The true function Z_i corresponding to argument X_i is approximated by the linear interpolate Z'_i.

To carry out a linear interpolation with a calculator, first compute $p = (X_i - X_1)(X_2 - X_1)$. Then substitute the given values of the function and p into the following equation:

$$Z'_i = pZ_2 + (1 - p)Z_1$$

The coefficients p and $1 - p$ represent complementary proportions of the distance from the tabled arguments to the intermediate value.

An example will show the use of this equation. Suppose we wish to find the value of the area of a normal curve corresponding to 2.1367 standard deviation units. In Table **A** we find arguments $X_1 = 2.13$ and $X_2 = 2.14$ and corresponding functions $Z_1 = 0.4834$ and $Z_2 = 0.4838$. We then compute the proportion $p = (2.1367 - 2.13) / (2.14 - 2.13) = (0.0067) / (0.01) = 0.67$. Then

$$Z'_i = (0.67)(0.4838) + (0.33)(0.4834) = (0.48367)$$

When, as frequently happens, the length of the interval from X_1 to X_2 is 1, the computation is especially simple, since $p = X_i - X_1$.

Inverse interpolation is employed to evaluate an argument given a value of a function intermediate between two tabled values. Using the same symbolism as above, one approximates the desired argument X_i by X'_i as follows:

$$X'_i = X_1 + \frac{(Z_i - Z_1)(X_2 - X_1)}{Z_2 - Z_1}$$

By way of illustration, interpolate for the argument in the earlier example from Table **A** where $Z'_i = 0.48367$ was obtained as the linearly interpolated value for $X_i = 2.1367$.

Bracketing this value Z_i are functions $Z_1 = 0.4834$ and $Z_2 = 0.4838$. The corresponding arguments in the table are $X_1 = 2.13$ and $X_2 = 2.14$. On substitution in the inverse interpolation formula, one obtains

$$X'_i = 2.13 + \frac{(0.48367 - 0.4834)(2.14 - 2.13)}{0.4838 - 0.4834}$$

$$= 2.13 + \frac{(0.00027)(0.01)}{0.0004}$$

$$= 2.13 + 0.00675$$

$$= 2.1368$$

Many tables, such as Table **B**, are arranged for *harmonic interpolation*. For the upper range of the arguments, functions in these tables will be approximately linearly related to the reciprocal of the arguments. Usually the arguments are degrees of freedom spaced as follows: 30, 40, 60, 120, ∞. For purposes of convenience in interpolation these are changed by dividing them into the last finite value of the argument yielding 120/30, 120/40, 120/60, 120/120, 120/∞, or 4, 3, 2, 1, 0. These integral values are the new arguments; functions for any argument between these tabled values are linearly interpolated between these transformed values. One advantage of this is that it permits interpolation between a finite and an infinite argument. An example will illustrate the method. Find the value of $t_{.001[200]}$. In Table **B** can be found $Z_1 = t_{.001[120]} = 3.373$ and $Z_2 = t_{.001[\infty]} = 3.291$. Transform the arguments $X_1 = 120$ and $X_2 = \infty$ to $X_1 = 120/120 = 1$ and $X_2 = 120/\infty = 0$. Transform X_i to 120/200 = 0.6, and apply the formula for linear interpolation:

$$p = \frac{0.6 - 1}{0 - 1} = 0.4$$

$$Z'_i = (0.4)(3.291) + (0.6)(3.373)$$

$$= 3.3402$$

which is rounded to 3.340.

Statistical Tables

TABLE **A** Areas of the normal curve

Arguments are furnished for the right half of the normal curve from the mean up to 3.5 standard deviations in increments of 0.01 standard deviation units, and from there to 4.9 standard deviations from the mean in increments of 0.1 standard deviation units. The quantity given is the area under the standard normal density function between the mean and the critical point. The area is generally labeled $\frac{1}{2} - \alpha$ (as shown in the figure). The proportion of the area between the mean and the critical value is given to four significant decimal places up to 3.49 standard deviations, and from there to 4.9 standard deviations to six significant decimal places.

The area between the mean and 2.21 standard deviations, for example, according to the table comprises 0.4864 or 48.64% of the total area of the curve. From this value one can easily compute two other quantities. The area beyond 2.21 standard deviations is the 0.5-complement of this function: $0.5000 - 0.4864 = 0.0136$, which is α, the proportion of the area of the curve to the right of 2.21 standard deviations. To find $1 - \alpha$, the total area of the curve to the left of 2.21, add 0.5 (the entire area of the left half of the curve) to the area given as a function in the table. Thus, the area to the left of 2.21 standard deviations is 0.9864.

By inverse use of the table and interpolation one can find the number of standard deviations corresponding to a given area. Thus, the value 0.3264 represents the area between the mean and 0.94 standard deviations.

This table has many applications in statistics. It is employed whenever one makes probability statements about normally distributed variables (Section 7.9), when normalizing a frequency distribution (Section 6.5), and in a variety of other applications.

The table was generated by using the polynomial expression given in C. Hastings, *Approximations for Digital Computers* (Princeton University Press, 1955) and shown below, which approximates the integral of the normal probability distribution with a maximum error of 7.5×10^{-8}:

$$P(X) - \tfrac{1}{2} = \tfrac{1}{2} - \frac{1}{(\sqrt{2\pi}e^{-X^2/2}t)(b_1 + t\{b_2 + t[b_3 + t(b_4 + tb_5)]\})}$$

where $t = 1/(1 + pX)$, $p = 0.2316419$, $b_1 = 0.319381530$, $b_2 = -0.356563782$, $b_3 = 1.781477937$, $b_4 = -1.821255978$, and $b_5 = 1.330274429$.

TABLE A Areas of the normal curve

$\gtrless$ Standard deviation units	0.	0.01	0.02	0.03	0.04
0.0	.0000	.0040	.0080	.0120	.0160
0.1	.0398	.0438	.0478	.0517	.0557
0.2	.0793	.0832	.0871	.0910	.0948
0.3	.1179	.1217	.1255	.1293	.1331
0.4	.1554	.1591	.1628	.1664	.1700
0.5	.1915	.1950	.1985	.2019	.2054
0.6	.2257	.2291	.2324	.2357	.2389
0.7	.2580	.2611	.2642	.2673	.2704
0.8	.2881	.2910	.2939	.2967	.2995
0.9	.3159	.3186	.3212	.3238	.3264
1.0	.3413	.3438	.3461	.3485	.3508
1.1	.3643	.3665	.3686	.3708	.3729
1.2	.3849	.3869	.3888	.3907	.3925
1.3	.4032	.4049	.4066	.4082	.4099
1.4	.4192	.4207	.4222	.4236	.4251
1.5	.4332	.4345	.4357	.4370	.4382
1.6	.4452	.4463	.4474	.4484	.4495
1.7	.4554	.4564	.4573	.4582	.4591
1.8	.4641	.4649	.4656	.4664	.4671
1.9	.4713	.4719	.4726	.4732	.4738
2.0	.4772	.4778	.4783	.4788	.4793
2.1	.4821	.4826	.4830	.4834	.4838
2.2	.4861	.4864	.4868	.4871	.4875
2.3	.4893	.4896	.4898	.4901	.4904
2.4	.4918	.4920	.4922	.4925	.4927
2.5	.4938	.4940	.4941	.4943	.4945
2.6	.4953	.4955	.4956	.4957	.4959
2.7	.4965	.4966	.4967	.4968	.4969
2.8	.4974	.4975	.4976	.4977	.4977
2.9	.4981	.4982	.4982	.4983	.4984
3.0	.4987	.4987	.4987	.4988	.4988
3.1	.4990	.4991	.4991	.4991	.4992
3.2	.4993	.4993	.4994	.4994	.4994
3.3	.4995	.4995	.4995	.4996	.4996
3.4	.4997	.4997	.4997	.4997	.4997
3.5	.499767				
3.6	.499841				
3.7	.499892				
3.8	.499928				
3.9	.499952				
4.0	.499968				
4.1	.499979				
4.2	.499987				
4.3	.499991				
4.4	.499995				
4.5	.499997				
4.6	.499998				
4.7	.499999				
4.8	.499999				
4.9	.500000				

0.05	0.06	0.07	0.08	0.09	Standard deviation units
.0199	.0239	.0279	.0319	.0359	0.0
.0596	.0636	.0675	.0714	.0753	0.1
.0987	.1026	.1064	.1103	.1141	0.2
.1368	.1406	.1443	.1480	.1517	0.3
.1736	.1772	.1808	.1844	.1879	0.4
.2088	.2123	.2157	.2190	.2224	0.5
.2422	.2454	.2486	.2517	.2549	0.6
.2734	.2764	.2794	.2823	.2852	0.7
.3023	.3051	.3078	.3106	.3133	0.8
.3289	.3315	.3340	.3365	.3389	0.9
.3531	.3554	.3577	.3599	.3621	1.0
.3749	.3770	.3790	.3810	.3830	1.1
.3944	.3962	.3980	.3997	.4015	1.2
.4115	.4131	.4147	.4162	.4177	1.3
.4265	.4279	.4292	.4306	.4319	1.4
.4394	.4406	.4418	.4429	.4441	1.5
.4505	.4515	.4525	.4535	.4545	1.6
.4599	.4608	.4616	.4625	.4633	1.7
.4678	.4686	.4693	.4699	.4706	1.8
.4744	.4750	.4756	.4761	.4767	1.9
.4798	.4803	.4808	.4812	.4817	2.0
.4842	.4846	.4850	.4854	.4857	2.1
.4878	.4881	.4884	.4887	.4890	2.2
.4906	.4909	.4911	.4913	.4916	2.3
.4929	.4931	.4932	.4934	.4936	2.4
.4946	.4948	.4949	.4951	.4952	2.5
.4960	.4961	.4962	.4963	.4964	2.6
.4970	.4971	.4972	.4973	.4974	2.7
.4978	.4979	.4979	.4980	.4981	2.8
.4984	.4985	.4985	.4986	.4986	2.9
.4989	.4989	.4989	.4990	.4990	3.0
.4992	.4992	.4992	.4993	.4993	3.1
.4994	.4994	.4995	.4995	.4995	3.2
.4996	.4996	.4996	.4996	.4997	3.3
.4997	.4997	.4997	.4997	.4998	3.4

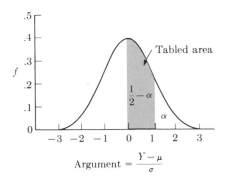

Tabled area

$\frac{1}{2} - \alpha$

α

$$\text{Argument} = \frac{Y - \mu}{\sigma}$$

$$t_s = \frac{\bar{X} - \mu_{\bar{x}}}{s_{\bar{x}}} \qquad s_{\bar{x}} = \frac{s_x^2}{n}$$

TABLE B Critical values of Student's t-distribution

This table furnishes critical values for Student's t-distribution for degrees of freedom $v = 1$ to 30 in increments of one, and for $v = 40$, 60, 120, and ∞. The percentage points (corresponding to $\alpha = 0.9$, 0.5, 0.4, 0.2, 0.1, 0.05, 0.02, 0.01, and 0.001) represent the area beyond the critical values of $\pm t$ in *both* tails of the distribution, as shown in the accompanying figure. The critical values of t are given to three decimal places.

To find the critical values of t for a given number of degrees of freedom, look up v in the left (argument) column of the table and read off the desired values of t in the corresponding row. For example, for 22 degrees of freedom, $t_{.05} = 2.074$ and $t_{.01} = 2.819$. The last value indicates that 1% of the area of the t-distribution (with 22 degrees of freedom) is beyond $t = \pm 2.819$, with 0.5% in each tail. If a one-tailed test is desired, the probabilities at the head of the table must be halved. Thus, for a one-tailed test with 4 *df*, the critical value $t = 3.747$ delimits 0.01 of the area of the curve instead of 0.02. For degrees of freedom $v > 30$, we need to interpolate between the values of the argument v. The table is designed for harmonic interpolation. Thus, to obtain $t_{.05[43]}$, interpolate between $t_{.05[40]} = 2.021$ and $t_{.05[60]} = 2.000$, which are furnished in the table. Transform the argument into $120/v = 120/43 = 2.791$ and interpolate between $120/60 = 2.000$ and $120/40 = 3.000$ by ordinary linear interpolation:

$$t_{.05[43]} = (0.791 \times 2.021) + [(1 - 0.791) \times 2.000]$$
$$= 2.017$$

When $v > 120$, interpolate between $120/\infty = 0$ and $120/120 = 1$. If critical values other than those furnished in this table are desired. E. S. Pearson and H. O. Hartley, *Biometrika Tables for Statisticians*, Vol. I (Cambridge University Press, 1958) quote J. B. Simaika (*Biometrika* 32:263–276, 1942), who recommends linear interpolation using the logarithms of the two-tailed probability values.

There are numerous applications of the t-distribution (Section 7.4) in statistics. Among them are the setting of confidence limits to means of small samples (Section 7.5), tests of difference between two means (Section 9.4), and comparisons of a single specimen with a sample (Section 9.5).

Values in this table have been taken from a more extensive one (table III) in R. A. Fisher and F. Yates, *Statistical Tables for Biological, Agricultural and Medical Research*, 5th ed. (Oliver & Boyd, Edinburgh, 1958) with permission of the authors and their publisher.

TABLE B   Critical values of Student's t-distribution

ν \ α	0.9	0.5	0.4	0.2	0.1	0.05	0.02	0.01	0.001	α / ν
1	.158	1.000	1.376	3.078	6.314	12.706	31.821	63.657	636.619	1
2	.142	.816	1.061	1.886	2.920	4.303	6.965	9.925	31.598	2
3	.137	.765	.978	1.638	2.353	3.182	4.541	5.841	12.924	3
4	.134	.741	.941	1.533	2.132	2.776	3.747	4.604	8.610	4
5	.132	.727	.920	1.476	2.015	2.571	3.365	4.032	6.869	5
6	.131	.718	.906	1.440	1.943	2.447	3.143	3.707	5.959	6
7	.130	.711	.896	1.415	1.895	2.365	2.998	3.499	5.408	7
8	.130	.706	.889	1.397	1.860	2.306	2.896	3.355	5.041	8
9	.129	.703	.883	1.383	1.833	2.262	2.821	3.250	4.781	9
10	.129	.700	.879	1.372	1.812	2.228	2.764	3.169	4.587	10
11	.129	.697	.876	1.363	1.796	2.201	2.718	3.106	4.437	11
12	.128	.695	.873	1.356	1.782	2.179	2.681	3.055	4.318	12
13	.128	.694	.870	1.350	1.771	2.160	2.650	3.012	4.221	13
14	.128	.692	.868	1.345	1.761	2.145	2.624	2.977	4.140	14
15	.128	.691	.866	1.341	1.753	2.131	2.602	2.947	4.073	15
16	.128	.690	.865	1.337	1.746	2.120	2.583	2.921	4.015	16
17	.128	.689	.863	1.333	1.740	2.110	2.567	2.898	3.965	17
18	.127	.688	.862	1.330	1.734	2.101	2.552	2.878	3.922	18
19	.127	.688	.861	1.328	1.729	2.093	2.539	2.861	3.883	19
20	.127	.687	.860	1.325	1.725	2.086	2.528	2.845	3.850	20
21	.127	.686	.859	1.323	1.721	2.080	2.518	2.831	3.819	21
22	.127	.686	.858	1.321	1.717	2.074	2.508	2.819	3.792	22
23	.127	.685	.858	1.319	1.714	2.069	2.500	2.807	3.767	23
24	.127	.685	.857	1.318	1.711	2.064	2.492	2.797	3.745	24
25	.127	.684	.856	1.316	1.708	2.060	2.485	2.787	3.725	25
26	.127	.684	.856	1.315	1.706	2.056	2.479	2.779	3.707	26
27	.127	.684	.855	1.314	1.703	2.052	2.473	2.771	3.690	27
28	.127	.683	.855	1.313	1.701	2.048	2.467	2.763	3.674	28
29	.127	.683	.854	1.311	1.699	2.045	2.462	2.756	3.659	29
30	.127	.683	.854	1.310	1.697	2.042	2.457	2.750	3.646	30
40	.126	.681	.851	1.303	1.684	2.021	2.423	2.704	3.551	40
60	.126	.679	.848	1.296	1.671	2.000	2.390	2.660	3.460	60
120	.126	.677	.845	1.289	1.658	1.980	2.358	2.617	3.373	120
∞	.126	.674	.842	1.282	1.645	1.960	2.326	2.576	3.291	∞

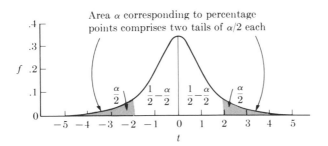

Area α corresponding to percentage points comprises two tails of $\alpha/2$ each

TABLE C Critical values of Student's *t*-distribution based on Šidák's multiplicative inequality

This table furnishes critical values for Student's *t*-distribution for unusual percentage points not found in standard *t*-tables such as Table **B**. These unusual percentage points α' are obtained from Šidák's multiplicative inequality: $\alpha' = 1 - (1 - \alpha)^{1/k}$, where α is the experimentwise error rate (also known as the familywise type I error rate) and k is the number of comparisons intended. Values of $t_{\alpha'}$ are furnished for degrees of freedom v from 2 to 24 in increments of one, for $v = 26, 28, 30, 40, 60, 120$, and ∞. These values of v lend themselves to harmonic interpolation as shown for Table **B**. Arguments for k, the number of comparisons, are given from 1 to 60 in increments of one. Three levels of experimentwise α (0.10, 0.05, and 0.01) are given for each combination of v and k in the table.

To find the critical value for a 5% experimentwise error based on a mean square with $v = 20$ degrees of freedom and making $k = 5$ comparisons overall, we enter the table at $k = 5$, $v = 20$, and $\alpha = .05$ to obtain $t_{.010206[20]} = 2.836$. This table is two-tailed. For a one-tailed test halve the experimentwise error rate α.

This table is employed for nonorthogonal multiple-comparisons tests (Section 9.6) and for setting multiple confidence intervals (Section 14.10).

The table was computed directly using a modification of the Hewlett Packard 9830 Statistical Distributions Pack, volume 1. All computations were carried out to 12 decimal places. This table extends that published by P. A. Games (*J. Amer. Stat. Assn.* **72**:531–534, 1977) from $k = 50$ to $k = 60$. The computations were checked by comparison with values published by Games. A few values differ by 1 in the last decimal place because of rounding error.

TABLE C Critical values of Student's t-distribution based on
Šidák's multiplicative inequality

Degrees of freedom ν

k	α	2	3	4	5	6	7	8	9	10	11
1	.1	2.920	2.353	2.132	2.015	1.943	1.895	1.860	1.833	1.812	1.796
	.05	4.303	3.182	2.776	2.571	2.447	2.365	2.306	2.262	2.228	2.201
	.01	9.925	5.841	4.604	4.032	3.707	3.499	3.355	3.250	3.169	3.106
2	.1	4.243	3.149	2.751	2.549	2.428	2.347	2.289	2.246	2.213	2.186
	.05	6.164	4.156	3.481	3.152	2.959	2.832	2.743	2.677	2.626	2.586
	.01	14.071	7.447	5.594	4.771	4.315	4.027	3.831	3.688	3.580	3.495
3	.1	5.243	3.690	3.150	2.882	2.723	2.618	2.544	2.488	2.446	2.412
	.05	7.582	4.826	3.941	3.518	3.274	3.115	3.005	2.923	2.860	2.811
	.01	17.248	8.565	6.248	5.243	4.695	4.353	4.120	3.952	3.825	3.726
4	.1	6.081	4.115	3.452	3.129	2.939	2.814	2.726	2.661	2.611	2.571
	.05	8.774	5.355	4.290	3.791	3.505	3.321	3.193	3.099	3.027	2.970
	.01	19.925	9.453	6.751	5.599	4.977	4.591	4.331	4.143	4.002	3.892
5	.1	6.816	4.471	3.699	3.327	3.110	2.969	2.869	2.796	2.739	2.695
	.05	9.823	5.799	4.577	4.012	3.690	3.484	3.342	3.237	3.157	3.094
	.01	22.282	10.201	7.166	5.888	5.203	4.782	4.498	4.294	4.141	4.022
6	.1	7.480	4.780	3.909	3.493	3.253	3.097	2.987	2.907	2.845	2.796
	.05	10.769	6.185	4.822	4.197	3.845	3.620	3.464	3.351	3.264	3.196
	.01	24.413	10.853	7.520	6.133	5.394	4.941	4.637	4.419	4.256	4.129
7	.1	8.090	5.055	4.093	3.638	3.376	3.206	3.088	3.001	2.934	2.881
	.05	11.639	6.529	5.036	4.358	3.978	3.736	3.569	3.448	3.355	3.283
	.01	26.372	11.436	7.832	6.346	5.559	5.078	4.756	4.526	4.354	4.221
8	.1	8.656	5.304	4.257	3.765	3.484	3.302	3.176	3.083	3.012	2.955
	.05	12.449	6.842	5.228	4.501	4.095	3.838	3.661	3.532	3.434	3.358
	.01	28.196	11.966	8.112	6.535	5.704	5.198	4.860	4.619	4.439	4.300
9	.1	9.188	5.532	4.406	3.880	3.580	3.388	3.254	3.155	3.080	3.021
	.05	13.208	7.128	5.402	4.630	4.200	3.929	3.743	3.607	3.505	3.424
	.01	29.908	12.453	8.367	6.706	5.835	5.306	4.953	4.703	4.515	4.371
10	.1	9.691	5.744	4.542	3.985	3.668	3.465	3.324	3.221	3.142	3.079
	.05	13.927	7.394	5.562	4.747	4.296	4.011	3.816	3.675	3.568	3.484
	.01	31.528	12.904	8.600	6.862	5.954	5.404	5.038	4.778	4.584	4.434
11	.1	10.169	5.941	4.668	4.081	3.748	3.535	3.388	3.280	3.197	3.133
	.05	14.610	7.642	5.710	4.855	4.383	4.086	3.883	3.736	3.625	3.538
	.01	33.068	13.326	8.817	7.006	6.063	5.493	5.115	4.846	4.646	4.492
12	.1	10.625	6.127	4.785	4.170	3.822	3.600	3.446	3.334	3.249	3.181
	.05	15.263	7.876	5.848	4.955	4.464	4.156	3.945	3.793	3.677	3.587
	.01	34.540	13.724	9.019	7.139	6.164	5.575	5.185	4.909	4.703	4.545
13	.1	11.063	6.302	4.895	4.253	3.891	3.660	3.501	3.384	3.296	3.226
	.05	15.889	8.096	5.977	5.049	4.539	4.220	4.002	3.845	3.726	3.633
	.01	35.951	14.099	9.208	7.264	6.258	5.652	5.251	4.967	4.756	4.594
14	.1	11.484	6.468	4.999	4.330	3.955	3.716	3.551	3.431	3.339	3.268
	.05	16.491	8.306	6.099	5.136	4.609	4.280	4.055	3.893	3.771	3.675
	.01	37.309	14.456	9.387	7.381	6.345	5.724	5.312	5.021	4.806	4.639
15	.1	11.890	6.627	5.097	4.403	4.015	3.768	3.598	3.474	3.380	3.306
	.05	17.072	8.505	6.214	5.219	4.675	4.336	4.105	3.939	3.813	3.715
	.01	38.620	14.796	9.556	7.491	6.428	5.791	5.370	5.072	4.852	4.682

TABLE C Critical values of Student's *t*-distribution based on Šidák's multiplicative inequality (*continued*)

Degrees of freedom *v*

k	α	2	3	4	5	6	7	8	9	10	11
16	.1	12.283	6.778	5.189	4.473	4.072	3.818	3.643	3.515	3.419	3.343
	.05	17.633	8.696	6.323	5.297	4.737	4.389	4.152	3.981	3.852	3.752
	.01	39.887	15.121	9.717	7.596	6.506	5.854	5.424	5.120	4.895	4.721
17	.1	12.663	6.923	5.278	4.538	4.125	3.864	3.685	3.554	3.455	3.377
	.05	18.178	8.879	6.428	5.371	4.796	4.439	4.196	4.021	3.889	3.787
	.01	41.116	15.433	9.871	7.695	6.580	5.914	5.475	5.165	4.936	4.759
18	.1	13.032	7.062	5.363	4.601	4.176	3.908	3.724	3.590	3.489	3.409
	.05	18.706	9.054	6.527	5.441	4.852	4.486	4.238	4.059	3.924	3.820
	.01	42.308	15.733	10.018	7.790	6.651	5.971	5.523	5.208	4.974	4.795
19	.1	13.392	7.196	5.444	4.660	4.225	3.950	3.762	3.625	3.521	3.439
	.05	19.220	9.223	6.623	5.509	4.905	4.531	4.278	4.095	3.958	3.851
	.01	43.468	16.021	10.158	7.881	6.718	6.025	5.569	5.248	5.011	4.828
20	.1	13.741	7.326	5.521	4.718	4.272	3.990	3.798	3.658	3.552	3.468
	.05	19.721	9.387	6.714	5.573	4.956	4.574	4.316	4.130	3.989	3.880
	.01	44.598	16.300	10.294	7.968	6.782	6.077	5.613	5.287	5.046	4.860
21	.1	14.083	7.451	5.596	4.772	4.316	4.029	3.832	3.689	3.581	3.496
	.05	20.209	9.544	6.803	5.635	5.005	4.615	4.352	4.162	4.020	3.909
	.01	45.700	16.570	10.424	8.051	6.844	6.126	5.655	5.324	5.079	4.891
22	.1	14.416	7.572	5.668	4.825	4.359	4.065	3.865	3.719	3.609	3.522
	.05	20.686	9.697	6.888	5.695	5.052	4.655	4.386	4.194	4.049	3.936
	.01	46.776	16.831	10.549	8.131	6.903	6.174	5.696	5.359	5.111	4.920
23	.1	14.741	7.689	5.738	4.876	4.400	4.100	3.896	3.748	3.636	3.548
	.05	21.152	9.846	6.970	5.752	5.097	4.693	4.420	4.224	4.076	3.962
	.01	47.828	17.084	10.671	8.208	6.960	6.220	5.734	5.393	5.141	4.948
24	.1	15.060	7.803	5.805	4.925	4.439	4.134	3.926	3.775	3.661	3.572
	.05	21.608	9.990	7.050	5.808	5.141	4.729	4.452	4.252	4.103	3.986
	.01	48.857	17.330	10.788	8.283	7.015	6.263	5.771	5.426	5.171	4.975
25	.1	15.371	7.914	5.870	4.972	4.477	4.167	3.955	3.802	3.686	3.595
	.05	22.054	10.130	7.127	5.861	5.182	4.764	4.482	4.280	4.128	4.010
	.01	49.865	17.569	10.902	8.355	7.068	6.306	5.807	5.457	5.199	5.001
26	.1	15.677	8.022	5.934	5.017	4.514	4.198	3.983	3.827	3.710	3.617
	.05	22.492	10.266	7.201	5.913	5.223	4.798	4.512	4.307	4.153	4.033
	.01	50.853	17.802	11.012	8.425	7.119	6.347	5.842	5.487	5.226	5.026
27	.1	15.977	8.127	5.995	5.062	4.549	4.229	4.010	3.852	3.732	3.639
	.05	22.921	10.399	7.274	5.963	5.262	4.831	4.540	4.333	4.177	4.055
	.01	51.822	18.029	11.120	8.492	7.168	6.386	5.875	5.516	5.252	5.050
28	.1	16.271	8.230	6.055	5.105	4.584	4.258	4.036	3.876	3.755	3.660
	.05	23.343	10.528	7.344	6.012	5.300	4.862	4.568	4.357	4.199	4.077
	.01	52.773	18.251	11.224	8.558	7.216	6.424	5.907	5.545	5.278	5.073
29	.1	16.560	8.330	6.113	5.146	4.617	4.286	4.061	3.899	3.776	3.680
	.05	23.757	10.654	7.413	6.059	5.336	4.893	4.595	4.382	4.222	4.097
	.01	53.707	18.467	11.325	8.622	7.263	6.461	5.938	5.572	5.302	5.095
30	.1	16.845	8.427	6.169	5.187	4.649	4.314	4.086	3.921	3.796	3.699
	.05	24.163	10.778	7.480	6.105	5.372	4.923	4.621	4.405	4.243	4.117
	.01	54.626	18.678	11.424	8.684	7.308	6.497	5.969	5.598	5.326	5.117

TABLE C Critical values of Student's *t*-distribution based on
Šidák's multiplicative inequality (*continued*)

Degrees of freedom v

k	α	2	3	4	5	6	7	8	9	10	11
31	.1	17.124	8.523	6.224	5.226	4.681	4.341	4.110	3.943	3.816	3.718
	.05	24.564	10.899	7.545	6.149	5.407	4.952	4.646	4.427	4.264	4.137
	.01	55.529	18.885	11.521	8.744	7.352	6.532	5.998	5.624	5.349	5.138
32	.1	17.399	8.616	6.278	5.264	4.711	4.367	4.133	3.963	3.836	3.736
	.05	24.957	11.017	7.608	6.193	5.440	4.980	4.670	4.449	4.284	4.155
	.01	56.418	19.087	11.615	8.803	7.395	6.566	6.026	5.649	5.371	5.159
33	.1	17.670	8.708	6.330	5.302	4.741	4.392	4.155	3.984	3.855	3.754
	.05	25.345	11.133	7.670	6.235	5.473	5.007	4.694	4.471	4.303	4.174
	.01	57.293	19.285	11.707	8.860	7.436	6.599	6.054	5.673	5.393	5.178
34	.1	17.936	8.798	6.381	5.338	4.770	4.417	4.177	4.004	3.873	3.771
	.05	25.727	11.246	7.731	6.277	5.505	5.033	4.717	4.491	4.322	4.191
	.01	58.155	19.479	11.796	8.916	7.477	6.631	6.081	5.697	5.414	5.198
35	.1	18.199	8.886	6.432	5.374	4.798	4.441	4.198	4.023	3.891	3.788
	.05	26.103	11.357	7.790	6.317	5.536	5.059	4.740	4.512	4.341	4.209
	.01	59.004	19.670	11.884	8.971	7.516	6.663	6.107	5.720	5.434	5.216
36	.1	18.458	8.972	6.480	5.408	4.826	4.464	4.218	4.041	3.908	3.804
	.05	26.474	11.466	7.848	6.356	5.567	5.085	4.762	4.531	4.359	4.225
	.01	59.841	19.856	11.970	9.024	7.555	6.693	6.133	5.742	5.454	5.235
37	.1	18.713	9.056	6.528	5.442	4.853	4.487	4.238	4.060	3.925	3.820
	.05	26.839	11.573	7.905	6.395	5.596	5.109	4.783	4.550	4.376	4.242
	.01	60.667	20.040	12.054	9.076	7.593	6.723	6.158	5.764	5.474	5.252
38	.1	18.965	9.140	6.575	5.475	4.879	4.509	4.258	4.077	3.941	3.835
	.05	27.200	11.678	7.961	6.433	5.625	5.133	4.804	4.569	4.393	4.257
	.01	61.481	20.220	12.137	9.127	7.630	6.752	6.182	5.785	5.493	5.270
39	.1	19.213	9.221	6.621	5.508	4.905	4.531	4.277	4.095	3.957	3.850
	.05	27.556	11.782	8.015	6.470	5.654	5.157	4.824	4.587	4.410	4.273
	.01	62.285	20.396	12.218	9.177	7.666	6.781	6.206	5.806	5.511	5.287
40	.1	19.459	9.301	6.667	5.540	4.930	4.552	4.296	4.112	3.973	3.865
	.05	27.908	11.883	8.069	6.506	5.682	5.180	4.844	4.605	4.426	4.288
	.01	63.079	20.570	12.297	9.226	7.701	6.809	6.230	5.826	5.529	5.303
41	.1	19.701	9.380	6.711	5.571	4.954	4.573	4.314	4.128	3.988	3.879
	.05	28.255	11.983	8.121	6.541	5.709	5.202	4.863	4.622	4.442	4.303
	.01	63.863	20.741	12.375	9.274	7.735	6.836	6.252	5.846	5.547	5.319
42	.1	19.941	9.458	6.754	5.601	4.979	4.593	4.332	4.144	4.003	3.893
	.05	28.598	12.081	8.173	6.576	5.735	5.224	4.882	4.639	4.458	4.317
	.01	64.637	20.909	12.451	9.321	7.769	6.863	6.275	5.865	5.564	5.335
43	.1	20.177	9.534	6.797	5.631	5.002	4.613	4.349	4.160	4.018	3.907
	.05	28.936	12.178	8.223	6.610	5.761	5.245	4.901	4.656	4.473	4.331
	.01	65.402	21.075	12.526	9.367	7.802	6.889	6.296	5.884	5.581	5.350
44	.1	20.411	9.609	6.839	5.661	5.025	4.632	4.366	4.176	4.032	3.920
	.05	29.271	12.273	8.273	6.643	5.787	5.266	4.919	4.672	4.488	4.345
	.01	66.159	21.238	12.600	9.412	7.835	6.914	6.318	5.902	5.598	5.365
45	.1	20.642	9.683	6.880	5.689	5.048	4.651	4.383	4.191	4.046	3.933
	.05	29.603	12.366	8.322	6.676	5.812	5.287	4.937	4.688	4.502	4.358
	.01	66.906	21.398	12.672	9.457	7.867	6.939	6.339	5.921	5.614	5.380

TABLE C   Critical values of Student's *t*-distribution based on
Šidák's multiplicative inequality (*continued*)

Degrees of freedom *v*

k	α	2	3	4	5	6	7	8	9	10	11
46	.1	20.871	9.756	6.921	5.718	5.070	4.670	4.400	4.206	4.060	3.946
	.05	29.930	12.459	8.370	6.708	5.836	5.307	4.955	4.704	4.516	4.372
	.01	67.646	21.557	12.743	9.500	7.898	6.964	6.359	5.938	5.630	5.395
47	.1	21.097	9.828	6.960	5.746	5.092	4.688	4.416	4.220	4.073	3.959
	.05	30.254	12.550	8.417	6.740	5.861	5.327	4.972	4.719	4.530	4.385
	.01	68.377	21.712	12.814	9.543	7.929	6.988	6.379	5.956	5.645	5.409
48	.1	21.321	9.899	7.000	5.773	5.113	4.706	4.432	4.234	4.086	3.971
	.05	30.574	12.639	8.463	6.771	5.884	5.347	4.989	4.734	4.544	4.397
	.01	69.101	21.866	12.883	9.585	7.959	7.012	6.399	5.973	5.661	5.423
49	.1	21.542	9.969	7.038	5.800	5.134	4.724	4.447	4.248	4.099	3.983
	.05	30.892	12.728	8.509	6.801	5.908	5.366	5.005	4.749	4.557	4.410
	.01	69.817	22.018	12.950	9.627	7.989	7.035	6.418	5.990	5.676	5.436
50	.1	21.761	10.038	7.076	5.826	5.155	4.741	4.462	4.262	4.112	3.995
	.05	31.206	12.815	8.554	6.831	5.930	5.385	5.021	4.763	4.571	4.422
	.01	70.526	22.167	13.017	9.668	8.018	7.058	6.437	6.006	5.690	5.450
51	.1	21.978	10.106	7.114	5.852	5.175	4.758	4.477	4.275	4.124	4.006
	.05	31.516	12.901	8.599	6.861	5.953	5.403	5.037	4.777	4.583	4.434
	.01	71.228	22.315	13.083	9.708	8.046	7.080	6.456	6.022	5.705	5.463
52	.1	22.193	10.173	7.150	5.878	5.195	4.775	4.492	4.289	4.136	4.018
	.05	31.824	12.986	8.642	6.890	5.975	5.421	5.053	4.791	4.596	4.446
	.01	71.923	22.460	13.148	9.747	8.074	7.102	6.474	6.038	5.719	5.475
53	.1	22.406	10.239	7.187	5.903	5.215	4.791	4.506	4.302	4.148	4.029
	.05	32.129	13.070	8.685	6.919	5.997	5.439	5.068	4.805	4.609	4.457
	.01	72.612	22.604	13.212	9.786	8.102	7.124	6.492	6.054	5.733	5.488
54	.1	22.617	10.304	7.222	5.928	5.234	4.808	4.520	4.314	4.160	4.040
	.05	32.431	13.153	8.728	6.947	6.018	5.456	5.083	4.818	4.621	4.468
	.01	73.294	22.746	13.275	9.824	8.129	7.145	6.510	6.069	5.746	5.500
55	.1	22.826	10.369	7.258	5.952	5.253	4.823	4.534	4.327	4.171	4.051
	.05	32.730	13.234	8.770	6.975	6.039	5.474	5.098	4.831	4.633	4.479
	.01	73.969	22.886	13.337	9.862	8.156	7.166	6.528	6.084	5.760	5.513
56	.1	23.033	10.433	7.292	5.976	5.272	4.839	4.548	4.339	4.183	4.061
	.05	33.027	13.315	8.811	7.002	6.060	5.491	5.113	4.844	4.645	4.490
	.01	74.639	23.025	13.399	9.899	8.183	7.187	6.545	6.099	5.773	5.525
57	.1	23.238	10.496	7.327	6.000	5.290	4.855	4.561	4.351	4.194	4.071
	.05	33.321	13.395	8.852	7.029	6.080	5.507	5.127	4.857	4.656	4.501
	.01	75.302	23.161	13.459	9.936	8.209	7.207	6.562	6.114	5.786	5.536
58	.1	23.441	10.558	7.361	6.023	5.309	4.870	4.575	4.363	4.205	4.082
	.05	33.612	13.474	8.892	7.056	6.100	5.524	5.141	4.870	4.668	4.512
	.01	75.960	23.297	13.519	9.972	8.235	7.228	6.578	6.128	5.799	5.548
59	.1	23.643	10.620	7.394	6.046	5.326	4.885	4.588	4.375	4.216	4.092
	.05	33.901	13.552	8.931	7.082	6.120	5.540	5.155	4.882	4.679	4.522
	.01	76.612	23.430	13.578	10.008	8.260	7.247	6.595	6.142	5.811	5.559
60	.1	23.843	10.681	7.427	6.069	5.344	4.899	4.600	4.386	4.226	4.102
	.05	34.187	13.629	8.971	7.108	6.140	5.556	5.169	4.894	4.690	4.532
	.01	77.259	23.563	13.636	10.043	8.285	7.267	6.611	6.156	5.824	5.571

TABLE C Critical values of Student's t-distribution based on Šidák's multiplicative inequality (continued)

Degrees of freedom ν

k	α	12	13	14	15	16	17	18	19	20	21
1	.1	1.782	1.771	1.761	1.753	1.746	1.740	1.734	1.729	1.725	1.721
	.05	2.179	2.160	2.145	2.131	2.120	2.110	2.101	2.093	2.086	2.080
	.01	3.055	3.012	2.977	2.947	2.921	2.898	2.878	2.861	2.845	2.831
2	.1	2.164	2.146	2.131	2.118	2.106	2.096	2.088	2.080	2.073	2.067
	.05	2.553	2.526	2.503	2.483	2.467	2.452	2.439	2.427	2.417	2.408
	.01	3.427	3.371	3.324	3.285	3.251	3.221	3.195	3.173	3.152	3.134
3	.1	2.384	2.361	2.342	2.325	2.311	2.298	2.287	2.277	2.269	2.261
	.05	2.770	2.737	2.709	2.685	2.665	2.647	2.631	2.617	2.605	2.594
	.01	3.647	3.582	3.528	3.482	3.443	3.409	3.379	3.353	3.329	3.308
4	.1	2.539	2.512	2.489	2.470	2.453	2.439	2.426	2.415	2.405	2.395
	.05	2.924	2.886	2.854	2.827	2.804	2.783	2.766	2.750	2.736	2.723
	.01	3.804	3.733	3.673	3.622	3.579	3.541	3.508	3.479	3.454	3.431
5	.1	2.658	2.628	2.603	2.582	2.563	2.547	2.532	2.520	2.508	2.498
	.05	3.044	3.002	2.967	2.937	2.911	2.889	2.869	2.852	2.836	2.822
	.01	3.927	3.850	3.785	3.731	3.684	3.644	3.609	3.578	3.550	3.525
6	.1	2.756	2.723	2.696	2.672	2.652	2.634	2.619	2.605	2.593	2.581
	.05	3.141	3.096	3.058	3.026	2.998	2.974	2.953	2.934	2.918	2.903
	.01	4.029	3.946	3.878	3.820	3.771	3.728	3.691	3.658	3.629	3.602
7	.1	2.838	2.803	2.774	2.748	2.726	2.708	2.691	2.676	2.663	2.651
	.05	3.224	3.176	3.135	3.101	3.072	3.046	3.024	3.004	2.986	2.970
	.01	4.114	4.028	3.956	3.895	3.844	3.799	3.760	3.725	3.695	3.667
8	.1	2.910	2.872	2.841	2.814	2.791	2.771	2.753	2.738	2.724	2.711
	.05	3.296	3.245	3.202	3.166	3.135	3.108	3.085	3.064	3.045	3.029
	.01	4.189	4.099	4.024	3.961	3.907	3.860	3.820	3.784	3.752	3.724
9	.1	2.973	2.933	2.900	2.872	2.848	2.826	2.808	2.791	2.777	2.764
	.05	3.359	3.306	3.261	3.224	3.191	3.163	3.138	3.116	3.097	3.080
	.01	4.256	4.162	4.084	4.019	3.963	3.914	3.872	3.835	3.802	3.773
10	.1	3.029	2.988	2.953	2.924	2.898	2.876	2.857	2.839	2.824	2.810
	.05	3.416	3.361	3.314	3.275	3.241	3.212	3.186	3.163	3.143	3.125
	.01	4.315	4.218	4.138	4.071	4.013	3.963	3.920	3.881	3.848	3.817
11	.1	3.080	3.037	3.001	2.970	2.944	2.921	2.901	2.883	2.867	2.853
	.05	3.468	3.410	3.362	3.321	3.286	3.256	3.229	3.206	3.185	3.166
	.01	4.369	4.270	4.187	4.118	4.058	4.007	3.962	3.923	3.888	3.857
12	.1	3.127	3.082	3.045	3.013	2.985	2.962	2.941	2.922	2.906	2.891
	.05	3.515	3.455	3.406	3.364	3.327	3.296	3.269	3.245	3.223	3.204
	.01	4.419	4.317	4.232	4.160	4.100	4.047	4.001	3.961	3.926	3.894
13	.1	3.170	3.124	3.085	3.052	3.024	2.999	2.977	2.958	2.941	2.926
	.05	3.558	3.497	3.446	3.402	3.365	3.333	3.305	3.280	3.258	3.238
	.01	4.465	4.360	4.273	4.200	4.138	4.084	4.037	3.996	3.960	3.927
14	.1	3.210	3.162	3.122	3.088	3.059	3.034	3.011	2.992	2.974	2.959
	.05	3.598	3.535	3.483	3.439	3.400	3.367	3.338	3.313	3.290	3.270
	.01	4.507	4.400	4.311	4.237	4.173	4.118	4.071	4.029	3.992	3.958
15	.1	3.247	3.198	3.157	3.122	3.092	3.066	3.043	3.023	3.005	2.989
	.05	3.636	3.571	3.518	3.472	3.433	3.399	3.370	3.343	3.320	3.300
	.01	4.547	4.438	4.347	4.271	4.206	4.150	4.102	4.059	4.021	3.987

TABLE C Critical values of Student's t-distribution based on Šidák's multiplicative inequality (*continued*)

Degrees of freedom v

k	α	12	13	14	15	16	17	18	19	20	21
16	.1	3.281	3.231	3.189	3.153	3.122	3.096	3.072	3.052	3.033	3.017
	.05	3.671	3.605	3.550	3.503	3.464	3.429	3.399	3.372	3.348	3.327
	.01	4.584	4.473	4.381	4.303	4.237	4.180	4.131	4.087	4.049	4.014
17	.1	3.314	3.262	3.219	3.183	3.151	3.124	3.100	3.079	3.060	3.043
	.05	3.704	3.637	3.581	3.533	3.492	3.457	3.426	3.399	3.375	3.353
	.01	4.619	4.506	4.412	4.333	4.266	4.208	4.158	4.114	4.075	4.040
18	.1	3.345	3.292	3.248	3.211	3.178	3.151	3.126	3.105	3.085	3.068
	.05	3.735	3.667	3.609	3.561	3.519	3.483	3.452	3.424	3.399	3.377
	.01	4.652	4.537	4.442	4.362	4.293	4.235	4.184	4.139	4.099	4.064
19	.1	3.374	3.320	3.275	3.237	3.204	3.176	3.151	3.129	3.109	3.092
	.05	3.765	3.695	3.637	3.587	3.545	3.508	3.476	3.448	3.423	3.400
	.01	4.684	4.567	4.470	4.389	4.319	4.260	4.208	4.162	4.122	4.086
20	.1	3.402	3.347	3.301	3.262	3.228	3.199	3.174	3.152	3.132	3.114
	.05	3.793	3.722	3.662	3.612	3.569	3.532	3.499	3.470	3.445	3.422
	.01	4.714	4.595	4.497	4.414	4.344	4.284	4.231	4.185	4.144	4.108
21	.1	3.428	3.372	3.325	3.286	3.252	3.222	3.196	3.173	3.153	3.135
	.05	3.820	3.747	3.687	3.636	3.592	3.554	3.521	3.492	3.466	3.443
	.01	4.742	4.622	4.522	4.439	4.368	4.306	4.253	4.206	4.165	4.128
22	.1	3.453	3.396	3.349	3.308	3.274	3.244	3.217	3.194	3.173	3.155
	.05	3.846	3.772	3.710	3.659	3.614	3.576	3.542	3.512	3.486	3.463
	.01	4.769	4.647	4.547	4.462	4.390	4.328	4.274	4.227	4.185	4.147
23	.1	3.477	3.419	3.371	3.330	3.295	3.264	3.237	3.214	3.193	3.174
	.05	3.870	3.795	3.733	3.680	3.635	3.596	3.562	3.532	3.505	3.481
	.01	4.795	4.672	4.570	4.484	4.411	4.349	4.294	4.246	4.204	4.166
24	.1	3.500	3.441	3.392	3.351	3.315	3.284	3.257	3.233	3.211	3.192
	.05	3.894	3.818	3.754	3.701	3.655	3.616	3.581	3.551	3.524	3.499
	.01	4.821	4.695	4.592	4.506	4.432	4.368	4.313	4.265	4.222	4.184
25	.1	3.522	3.463	3.413	3.370	3.334	3.303	3.275	3.251	3.229	3.210
	.05	3.916	3.839	3.775	3.721	3.675	3.634	3.599	3.569	3.541	3.517
	.01	4.845	4.718	4.614	4.526	4.451	4.387	4.332	4.283	4.239	4.201
26	.1	3.544	3.483	3.432	3.389	3.353	3.321	3.293	3.268	3.246	3.227
	.05	3.938	3.860	3.795	3.740	3.693	3.653	3.617	3.586	3.558	3.533
	.01	4.868	4.740	4.634	4.546	4.470	4.406	4.349	4.300	4.256	4.217
27	.1	3.564	3.503	3.451	3.408	3.371	3.338	3.310	3.285	3.263	3.243
	.05	3.959	3.880	3.814	3.758	3.711	3.670	3.634	3.602	3.574	3.549
	.01	4.890	4.761	4.654	4.565	4.489	4.423	4.366	4.316	4.272	4.233
28	.1	3.584	3.522	3.470	3.426	3.388	3.355	3.326	3.301	3.279	3.259
	.05	3.979	3.899	3.832	3.776	3.728	3.687	3.650	3.618	3.590	3.565
	.01	4.912	4.781	4.673	4.583	4.506	4.440	4.383	4.332	4.288	4.248
29	.1	3.603	3.540	3.487	3.443	3.404	3.371	3.342	3.317	3.294	3.274
	.05	3.998	3.917	3.850	3.793	3.745	3.703	3.666	3.634	3.605	3.580
	.01	4.932	4.801	4.692	4.601	4.523	4.457	4.399	4.348	4.303	4.263
30	.1	3.621	3.557	3.504	3.459	3.420	3.387	3.358	3.332	3.309	3.288
	.05	4.017	3.935	3.867	3.810	3.761	3.718	3.681	3.649	3.620	3.594
	.01	4.953	4.819	4.710	4.618	4.540	4.472	4.414	4.363	4.317	4.277

TABLE **C** Critical values of Student's *t*-distribution based on
Šidák's multiplicative inequality (*continued*)

Degrees of freedom *v*

k	α	12	13	14	15	16	17	18	19	20	21
31	.1	3.639	3.575	3.521	3.475	3.436	3.402	3.372	3.346	3.323	3.302
	.05	4.035	3.953	3.884	3.826	3.776	3.733	3.696	3.663	3.634	3.608
	.01	4.972	4.838	4.727	4.634	4.556	4.488	4.429	4.377	4.331	4.291
32	.1	3.657	3.591	3.537	3.490	3.451	3.417	3.387	3.360	3.337	3.316
	.05	4.053	3.969	3.900	3.841	3.791	3.748	3.710	3.677	3.647	3.621
	.01	4.991	4.856	4.744	4.650	4.571	4.503	4.443	4.391	4.345	4.304
33	.1	3.673	3.607	3.552	3.505	3.465	3.431	3.401	3.374	3.350	3.329
	.05	4.070	3.986	3.916	3.857	3.806	3.762	3.724	3.690	3.661	3.634
	.01	5.009	4.873	4.760	4.666	4.586	4.517	4.457	4.405	4.358	4.317
34	.1	3.690	3.623	3.567	3.520	3.480	3.445	3.414	3.387	3.363	3.342
	.05	4.087	4.002	3.931	3.871	3.820	3.776	3.737	3.704	3.673	3.647
	.01	5.027	4.890	4.776	4.681	4.600	4.531	4.471	4.418	4.371	4.329
35	.1	3.705	3.638	3.582	3.534	3.493	3.458	3.427	3.400	3.376	3.354
	.05	4.103	4.017	3.946	3.885	3.834	3.789	3.750	3.716	3.686	3.659
	.01	5.045	4.906	4.792	4.696	4.614	4.544	4.484	4.430	4.383	4.342
36	.1	3.721	3.653	3.596	3.548	3.507	3.471	3.440	3.412	3.388	3.366
	.05	4.119	4.032	3.960	3.899	3.847	3.802	3.763	3.729	3.698	3.671
	.01	5.061	4.922	4.806	4.710	4.628	4.558	4.497	4.443	4.395	4.353
37	.1	3.736	3.667	3.610	3.561	3.520	3.484	3.452	3.424	3.400	3.378
	.05	4.134	4.047	3.974	3.913	3.860	3.815	3.775	3.741	3.710	3.682
	.01	5.078	4.937	4.821	4.724	4.641	4.571̵	4.509	4.455	4.407	4.365
38	.1	3.750	3.681	3.623	3.574	3.532	3.496	3.464	3.436	3.411	3.389
	.05	4.149	4.061	3.987	3.926	3.873	3.827	3.787	3.752	3.721	3.693
	.01	5.094	4.952	4.835	4.737	4.654	4.583	4.521	4.467	4.419	4.376
39	.1	3.765	3.695	3.636	3.587	3.544	3.508	3.476	3.448	3.422	3.400
	.05	4.164	4.075	⸱4.001	3.938	3.885	3.839	3.799	3.764	3.732	3.704
	.01	5.110	4.967	4.849	4.751	4.667	4.595	4.533	4.478	4.430	4.387
40	.1	3.779	3.708	3.649	3.599	3.556	3.519	3.487	3.459	3.433	3.411
	.05	4.178	4.088	4.014	3.951	3.897	3.851	3.810	3.775	3.743	3.715
	.01	5.125	4.981	4.863	4.764	4.679	4.607	4.544	4.489	4.441	4.397
41	.1	3.792	3.721	3.661	3.611	3.568	3.531	3.498	3.470	3.444	3.421
	.05	4.192	4.101	4.026	3.963	3.909	3.862	3.821	3.786	3.754	3.725
	.01	5.140	4.995	4.876	4.776	4.691	4.619	4.556	4.500	4.451	4.408
42	.1	3.805	3.733	3.673	3.623	3.579	3.542	3.509	3.480	3.454	3.432
	.05	4.205	4.114	4.038	3.975	3.920	3.873	3.832	3.796	3.764	3.736
	.01	5.154	5.009	4.889	4.788	4.703	4.630	4.566	4.511	4.462	4.418
43	.1	3.818	3.746	3.685	3.634	3.591	3.553	3.520	3.491	3.465	3.442
	.05	4.218	4.127	4.050	3.986	3.931	3.884	3.843	3.806	3.774	3.745
	.01	5.169	5.022	4.901	4.800	4.715	4.641	4.577	4.521	4.472	4.428
44	.1	3.831	3.758	3.697	3.646	3.601	3.563	3.530	3.501	3.475	3.451
	.05	4.231	4.139	4.062	3.998	3.942	3.895	3.853	3.816	3.784	3.755
	.01	5.183	5.035	4.914	4.812	4.726	4.652	4.588	4.531	4.482	4.437
45	.1	3.843	3.770	3.708	3.656	3.612	3.574	3.540	3.511	3.484	3.461
	.05	4.244	4.151	4.074	4.009	3.953	3.905	3.863⸱	3.826	3.794	3.765
	.01	5.196	5.048	4.926	4.824	4.737	4.662	4.598	4.541	4.491	4.447

TABLE C Critical values of Student's *t*-distribution based on
Šidák's multiplicative inequality (*continued*)

Degrees of freedom *v*

k	*α*	12	13	14	15	16	17	18	19	20	21
46	.1	3.855	3.781	3.719	3.667	3.623	3.584	3.550	3.520	3.494	3.470
	.05	4.256	4.163	4.085	4.019	3.963	3.915	3.873	3.836	3.803	3.774
	.01	5.209	5.060	4.938	4.835	4.748	4.673	4.608	4.551	4.501	4.456
47	.1	3.867	3.792	3.730	3.678	3.633	3.594	3.560	3.530	3.503	3.479
	.05	4.269	4.174	4.096	4.030	3.974	3.925	3.883	3.845	3.812	3.783
	.01	5.223	5.073	4.949	4.846	4.758	4.683	4.618	4.560	4.510	4.465
48	.1	3.879	3.804	3.741	3.688	3.643	3.603	3.569	3.539	3.512	3.488
	.05	4.281	4.186	4.107	4.040	3.984	3.935	3.892	3.855	3.821	3.792
	.01	5.235	5.084	4.960	4.857	4.768	4.693	4.627	4.570	4.519	4.474
49	.1	3.890	3.814	3.751	3.698	3.652	3.613	3.578	3.548	3.521	3.497
	.05	4.292	4.197	4.117	4.050	3.993	3.944	3.901	3.864	3.830	3.800
	.01	5.248	5.096	4.971	4.867	4.779	4.703	4.637	4.579	4.528	4.482
50	.1	3.901	3.825	3.761	3.708	3.662	3.622	3.587	3.557	3.530	3.505
	.05	4.304	4.207	4.128	4.060	4.003	3.954	3.910	3.872	3.839	3.809
	.01	5.260	5.108	4.982	4.877	4.788	4.712	4.646	4.588	4.536	4.491
51	.1	3.912	3.835	3.772	3.718	3.671	3.631	3.596	3.566	3.538	3.514
	.05	4.315	4.218	4.138	4.070	4.013	3.963	3.919	3.881	3.847	3.817
	.01	5.272	5.119	4.993	4.888	4.798	4.721	4.655	4.596	4.545	4.499
52	.1	3.923	3.846	3.781	3.727	3.680	3.640	3.605	3.574	3.547	3.522
	.05	4.326	4.228	4.148	4.080	4.022	3.972	3.928	3.890	3.856	3.825
	.01	5.284	5.130	5.004	4.898	4.808	4.731	4.664	4.605	4.553	4.507
53	.1	3.934	3.856	3.791	3.736	3.689	3.649	3.614	3.582	3.555	3.530
	.05	4.337	4.239	4.157	4.089	4.031	3.980	3.937	3.898	3.864	3.833
	.01	5.296	5.141	5.014	4.907	4.817	4.740	4.672	4.613	4.561	4.515
54	.1	3.944	3.866	3.800	3.745	3.698	3.658	3.622	3.591	3.563	3.538
	.05	4.347	4.249	4.167	4.098	4.040	3.989	3.945	3.906	3.872	3.841
	.01	5.307	5.152	5.024	4.917	4.826	4.748	4.681	4.622	4.569	4.523
55	.1	3.954	3.875	3.810	3.754	3.707	3.666	3.630	3.599	3.571	3.546
	.05	4.358	4.259	4.177	4.107	4.048	3.998	3.953	3.914	3.880	3.849
	.01	5.319	5.162	5.034	4.926	4.835	4.757	4.689	4.630	4.577	4.531
56	.1	3.964	3.885	3.819	3.763	3.716	3.674	3.638	3.607	3.579	3.553
	.05	4.368	4.268	4.186	4.116	4.057	4.006	3.961	3.922	3.887	3.856
	.01	5.330	5.173	5.044	4.936	4.844	4.766	4.697	4.638	4.585	4.538
57	.1	3.974	3.894	3.828	3.772	3.724	3.682	3.646	3.614	3.586	3.561
	.05	4.378	4.278	4.195	4.125	4.065	4.014	3.969	3.930	3.895	3.864
	.01	5.341	5.183	5.053	4.945	4.853	4.774	4.706	4.646	4.593	4.546
58	.1	3.983	3.903	3.837	3.780	3.732	3.690	3.654	3.622	3.594	3.568
	.05	4.388	4.287	4.204	4.134	4.074	4.022	3.977	3.937	3.902	3.871
	.01	5.351	5.193	5.063	4.954	4.861	4.782	4.713	4.653	4.600	4.553
59	.1	3.993	3.912	3.845	3.789	3.740	3.698	3.662	3.630	3.601	3.575
	.05	4.398	4.296	4.213	4.142	4.082	4.030	3.985	3.945	3.910	3.878
	.01	5.362	5.203	5.072	4.962	4.870	4.790	4.721	4.661	4.608	4.560
60	.1	4.002	3.921	3.854	3.797	3.748	3.706	3.669	3.637	3.608	3.583
	.05	4.407	4.305	4.221	4.150	4.090	4.038	3.992	3.952	3.917	3.885
	.01	5.372	5.212	5.081	4.971	4.878	4.798	4.729	4.668	4.615	4.567

18

TABLE C   Critical values of Student's *t*-distribution based on
Šidák's multiplicative inequality (*continued*)

Degrees of freedom v

k	α	22	23	24	26	28	30	40	60	120	∞
1	.1	1.717	1.714	1.711	1.706	1.701	1.697	1.684	1.671	1.658	1.645
	.05	2.074	2.069	2.064	2.056	2.048	2.042	2.021	2.000	1.980	1.960
	.01	2.819	2.807	2.797	2.779	2.763	2.750	2.704	2.660	2.617	2.576
2	.1	2.061	2.056	2.051	2.043	2.036	2.030	2.009	1.989	1.968	1.949
	.05	2.400	2.392	2.385	2.373	2.363	2.354	2.323	2.294	2.265	2.236
	.01	3.118	3.103	3.089	3.066	3.046	3.029	2.970	2.914	2.859	2.806
3	.1	2.254	2.247	2.241	2.231	2.222	2.215	2.189	2.163	2.138	2.114
	.05	2.584	2.574	2.566	2.551	2.539	2.528	2.492	2.456	2.422	2.388
	.01	3.289	3.272	3.257	3.230	3.207	3.188	3.121	3.056	2.994	2.934
4	.1	2.387	2.380	2.373	2.361	2.351	2.342	2.312	2.283	2.254	2.226
	.05	2.712	2.701	2.692	2.675	2.661	2.649	2.608	2.568	2.529	2.491
	.01	3.410	3.392	3.375	3.345	3.320	3.298	3.225	3.155	3.087	3.022
5	.1	2.489	2.481	2.473	2.460	2.449	2.439	2.406	2.373	2.342	2.311
	.05	2.810	2.798	2.788	2.770	2.755	2.742	2.696	2.653	2.610	2.569
	.01	3.503	3.483	3.465	3.433	3.407	3.384	3.305	3.230	3.158	3.089
6	.1	2.572	2.563	2.554	2.540	2.528	2.517	2.481	2.446	2.411	2.378
	.05	2.889	2.877	2.866	2.847	2.830	2.816	2.768	2.721	2.675	2.631
	.01	3.579	3.558	3.539	3.505	3.477	3.453	3.370	3.291	3.215	3.143
7	.1	2.641	2.631	2.622	2.607	2.594	2.582	2.544	2.506	2.469	2.434
	.05	2.956	2.943	2.931	2.911	2.893	2.878	2.827	2.777	2.729	2.683
	.01	3.643	3.621	3.601	3.566	3.536	3.511	3.425	3.342	3.263	3.188
8	.1	2.700	2.690	2.680	2.664	2.650	2.638	2.597	2.558	2.519	2.481
	.05	3.014	3.000	2.988	2.966	2.948	2.932	2.878	2.826	2.776	2.727
	.01	3.698	3.675	3.654	3.618	3.587	3.561	3.472	3.386	3.304	3.226
9	.1	2.752	2.741	2.731	2.714	2.700	2.687	2.644	2.603	2.562	2.523
	.05	3.064	3.050	3.037	3.014	2.995	2.979	2.923	2.869	2.816	2.766
	.01	3.747	3.723	3.702	3.664	3.632	3.605	3.513	3.425	3.340	3.260
10	.1	2.798	2.787	2.777	2.759	2.744	2.731	2.686	2.643	2.600	2.560
	.05	3.109	3.094	3.081	3.058	3.038	3.021	2.963	2.906	2.852	2.800
	.01	3.790	3.766	3.744	3.705	3.672	3.644	3.549	3.459	3.372	3.289
11	.1	2.840	2.828	2.818	2.799	2.783	2.770	2.723	2.678	2.635	2.592
	.05	3.150	3.134	3.121	3.096	3.076	3.058	2.998	2.940	2.884	2.830
	.01	3.830	3.804	3.782	3.742	3.708	3.680	3.582	3.489	3.401	3.316
12	.1	2.878	2.866	2.855	2.835	2.819	2.805	2.757	2.711	2.666	2.622
	.05	3.187	3.171	3.157	3.132	3.111	3.092	3.031	2.971	2.913	2.858
	.01	3.865	3.840	3.816	3.776	3.741	3.712	3.612	3.517	3.427	3.340
13	.1	2.912	2.900	2.889	2.869	2.852	2.838	2.788	2.740	2.694	2.649
	.05	3.220	3.204	3.190	3.164	3.142	3.124	3.060	2.999	2.940	2.883
	.01	3.898	3.872	3.848	3.807	3.771	3.742	3.640	3.543	3.451	3.362
14	.1	2.944	2.932	2.920	2.900	2.882	2.868	2.817	2.768	2.720	2.674
	.05	3.252	3.235	3.220	3.194	3.172	3.153	3.088	3.025	2.965	2.906
	.01	3.929	3.902	3.878	3.835	3.799	3.769	3.665	3.567	3.473	3.383
15	.1	2.974	2.961	2.949	2.928	2.911	2.895	2.843	2.793	2.744	2.697
	.05	3.281	3.264	3.249	3.222	3.199	3.180	3.113	3.049	2.987	2.928
	.01	3.957	3.930	3.905	3.862	3.825	3.794	3.689	3.589	3.493	3.402

TABLE C Critical values of Student's t-distribution based on
Šidák's multiplicative inequality (*continued*)

Degrees of freedom v

k	α	22	23	24	26	28	30	40	60	120	∞
16	.1	3.002	2.989	2.976	2.955	2.937	2.921	2.868	2.816	2.767	2.718
	.05	3.308	3.291	3.275	3.248	3.224	3.205	3.137	3.071	3.008	2.948
	.01	3.983	3.956	3.931	3.887	3.850	3.818	3.711	3.609	3.512	3.419
17	.1	3.028	3.014	3.002	2.980	2.961	2.946	2.891	2.838	2.787	2.738
	.05	3.333	3.316	3.300	3.272	3.248	3.228	3.159	3.092	3.028	2.966
	.01	4.008	3.980	3.955	3.910	3.872	3.840	3.732	3.628	3.530	3.436
18	.1	3.053	3.039	3.026	3.003	2.985	2.968	2.913	2.859	2.807	2.757
	.05	3.357	3.340	3.323	3.295	3.271	3.250	3.180	3.112	3.047	2.984
	.01	4.032	4.003	3.977	3.932	3.894	3.861	3.751	3.646	3.546	3.451
19	.1	3.076	3.061	3.048	3.026	3.006	2.990	2.933	2.878	2.826	2.774
	.05	3.380	3.362	3.345	3.316	3.292	3.271	3.199	3.130	3.064	3.000
	.01	4.054	4.025	3.999	3.953	3.914	3.881	3.769	3.663	3.562	3.466
20	.1	3.098	3.083	3.070	3.047	3.027	3.010	2.952	2.897	2.843	2.791
	.05	3.402	3.383	3.366	3.337	3.312	3.291	3.218	3.148	3.081	3.016
	.01	4.075	4.046	4.019	3.972	3.933	3.900	3.787	3.679	3.577	3.479
21	.1	3.118	3.104	3.090	3.067	3.047	3.029	2.971	2.914	2.860	2.807
	.05	3.422	3.403	3.386	3.356	3.331	3.309	3.235	3.165	3.096	3.031
	.01	4.095	4.065	4.038	3.991	3.952	3.918	3.803	3.694	3.591	3.493
22	.1	3.138	3.123	3.109	3.085	3.065	3.048	2.988	2.931	2.875	2.822
	.05	3.441	3.422	3.405	3.375	3.349	3.327	3.252	3.180	3.111	3.045
	.01	4.114	4.084	4.057	4.009	3.969	3.935	3.819	3.709	3.604	3.505
23	.1	3.157	3.142	3.128	3.104	3.083	3.065	3.005	2.947	2.890	2.836
	.05	3.460	3.441	3.423	3.392	3.366	3.344	3.268	3.195	3.125	3.058
	.01	4.132	4.102	4.074	4.026	3.985	3.951	3.834	3.723	3.617	3.517
24	.1	3.175	3.160	3.146	3.121	3.100	3.082	3.021	2.962	2.904	2.849
	.05	3.478	3.458	3.440	3.409	3.383	3.360	3.283	3.210	3.139	3.071
	.01	4.150	4.119	4.091	4.042	4.001	3.966	3.848	3.736	3.629	3.528
25	.1	3.193	3.177	3.162	3.137	3.116	3.098	3.036	2.976	2.918	2.862
	.05	3.495	3.475	3.457	3.425	3.399	3.376	3.298	3.223	3.152	3.083
	.01	4.166	4.135	4.107	4.058	4.017	3.981	3.862	3.749	3.641	3.539
26	.1	3.209	3.193	3.179	3.153	3.132	3.113	3.050	2.990	2.931	2.875
	.05	3.511	3.491	3.473	3.441	3.414	3.391	3.312	3.237	3.164	3.095
	.01	4.182	4.151	4.122	4.073	4.031	3.996	3.875	3.761	3.652	3.549
27	.1	3.225	3.209	3.194	3.169	3.147	3.128	3.064	3.003	2.944	2.887
	.05	3.527	3.506	3.488	3.456	3.428	3.405	3.326	3.249	3.176	3.106
	.01	4.198	4.166	4.137	4.087	4.045	4.009	3.888	3.773	3.663	3.559
28	.1	3.240	3.224	3.209	3.183	3.161	3.142	3.078	3.016	2.956	2.898
	.05	3.542	3.521	3.503	3.470	3.442	3.419	3.339	3.261	3.187	3.116
	.01	4.213	4.181	4.152	4.101	4.059	4.023	3.900	3.784	3.673	3.569
29	.1	3.255	3.239	3.224	3.197	3.175	3.156	3.091	3.028	2.967	2.909
	.05	3.557	3.536	3.517	3.484	3.456	3.432	3.351	3.273	3.198	3.127
	.01	4.227	4.195	4.166	4.115	4.072	4.035	3.912	3.795	3.683	3.578
30	.1	3.270	3.253	3.238	3.211	3.188	3.169	3.103	3.040	2.979	2.920
	.05	3.571	3.550	3.531	3.497	3.469	3.445	3.363	3.284	3.209	3.137
	.01	4.241	4.208	4.179	4.128	4.084	4.048	3.923	3.805	3.693	3.587

TABLE C Critical values of Student's *t*-distribution based on Šidák's multiplicative inequality (*continued*)

Degrees of freedom *ν*

k	α	22	23	24	26	28	30	40	60	120	∞
31	.1	3.283	3.266	3.251	3.224	3.201	3.182	3.115	3.051	2.989	2.930
	.05	3.584	3.563	3.544	3.510	3.482	3.457	3.375	3.295	3.219	3.146
	.01	4.254	4.222	4.192	4.140	4.097	4.060	3.934	3.815	3.702	3.595
32	.1	3.297	3.280	3.264	3.237	3.214	3.194	3.127	3.062	3.000	2.940
	.05	3.597	3.576	3.557	3.523	3.494	3.469	3.386	3.306	3.229	3.156
	.01	4.267	4.234	4.204	4.152	4.108	4.071	3.945	3.825	3.711	3.603
33	.1	3.310	3.292	3.277	3.249	3.226	3.206	3.138	3.073	3.010	2.949
	.05	3.610	3.589	3.569	3.535	3.506	3.481	3.397	3.316	3.239	3.165
	.01	4.280	4.247	4.216	4.164	4.120	4.082	3.955	3.834	3.720	3.611
34	.1	3.322	3.305	3.289	3.261	3.238	3.218	3.149	3.083	3.019	2.958
	.05	3.622	3.601	3.581	3.546	3.517	3.492	3.407	3.326	3.248	3.173
	.01	4.292	4.259	4.228	4.175	4.131	4.093	3.965	3.843	3.728	3.619
35	.1	3.334	3.317	3.301	3.273	3.249	3.229	3.160	3.093	3.029	2.967
	.05	3.634	3.613	3.593	3.558	3.528	3.503	3.418	3.336	3.257	3.182
	.01	4.304	4.270	4.240	4.186	4.142	4.103	3.975	3.852	3.736	3.627
36	.1	3.346	3.329	3.312	3.284	3.260	3.240	3.170	3.103	3.038	2.976
	.05	3.646	3.624	3.604	3.569	3.539	3.514	3.427	3.345	3.266	3.190
	.01	4.316	4.282	4.251	4.197	4.152	4.114	3.984	3.861	3.744	3.634
37	.1	3.358	3.340	3.324	3.295	3.271	3.250	3.180	3.112	3.047	2.984
	.05	3.657	3.635	3.615	3.579	3.550	3.524	3.437	3.354	3.274	3.198
	.01	4.327	4.293	4.262	4.208	4.162	4.124	3.993	3.869	3.752	3.641
38	.1	3.369	3.351	3.334	3.306	3.281	3.261	3.190	3.121	3.056	2.992
	.05	3.669	3.646	3.626	3.590	3.560	3.534	3.446	3.363	3.282	3.205
	.01	4.338	4.303	4.272	4.218	4.172	4.133	4.002	3.877	3.760	3.648
39	.1	3.380	3.362	3.345	3.316	3.292	3.271	3.199	3.130	3.064	3.000
	.05	3.679	3.657	3.636	3.600	3.570	3.544	3.456	3.371	3.290	3.213
	.01	4.348	4.314	4.282	4.228	4.182	4.143	4.010	3.885	3.767	3.655
40	.1	3.390	3.372	3.355	3.326	3.301	3.280	3.208	3.139	3.072	3.008
	.05	3.690	3.667	3.646	3.610	3.579	3.553	3.464	3.379	3.298	3.220
	.01	4.359	4.324	4.292	4.237	4.191	4.152	4.019	3.893	3.774	3.661
41	.1	3.401	3.382	3.365	3.336	3.311	3.290	3.217	3.147	3.080	3.015
	.05	3.700	3.677	3.656	3.620	3.589	3.563	3.473	3.388	3.306	3.227
	.01	4.369	4.334	4.302	4.247	4.200	4.161	4.027	3.901	3.781	3.667
42	.1	3.411	3.392	3.375	3.346	3.321	3.299	3.226	3.155	3.088	3.023
	.05	3.710	3.687	3.666	3.629	3.598	3.572	3.482	3.395	3.313	3.234
	.01	4.379	4.343	4.311	4.256	4.209	4.169	4.035	3.908	3.787	3.673
43	.1	3.421	3.402	3.385	3.355	3.330	3.308	3.234	3.163	3.095	3.030
	.05	3.720	3.696	3.675	3.638	3.607	3.580	3.490	3.403	3.320	3.241
	.01	4.388	4.353	4.321	4.265	4.218	4.178	4.043	3.915	3.794	3.679
44	.1	3.430	3.411	3.394	3.364	3.339	3.317	3.243	3.171	3.103	3.037
	.05	3.729	3.706	3.684	3.647	3.616	3.589	3.498	3.411	3.327	3.247
	.01	4.398	4.362	4.330	4.274	4.226	4.186	4.050	3.922	3.800	3.685
45	.1	3.440	3.421	3.403	3.373	3.347	3.325	3.251	3.179	3.110	3.043
	.05	3.738	3.715	3.693	3.656	3.624	3.597	3.506	3.418	3.334	3.254
	.01	4.407	4.371	4.339	4.282	4.235	4.194	4.058	3.929	3.807	3.691

TABLE C Critical values of Student's *t*-distribution based on Šidák's multiplicative inequality (*continued*)

Degrees of freedom *v*

k	α	22	23	24	26	28	30	40	60	120	∞
46	.1	3.449	3.430	3.412	3.382	3.356	3.334	3.259	3.186	3.117	3.050
	.05	3.747	3.724	3.702	3.664	3.633	3.606	3.513	3.425	3.341	3.260
	.01	4.416	4.380	4.347	4.290	4.243	4.202	4.065	3.935	3.813	3.697
47	.1	3.458	3.439	3.421	3.390	3.364	3.342	3.266	3.194	3.124	3.057
	.05	3.756	3.732	3.711	3.673	3.641	3.614	3.521	3.432	3.347	3.266
	.01	4.425	4.388	4.356	4.299	4.251	4.210	4.072	3.942	3.819	3.702
48	.1	3.467	3.447	3.429	3.399	3.372	3.350	3.274	3.201	3.130	3.063
	.05	3.765	3.741	3.719	3.681	3.649	3.621	3.528	3.439	3.354	3.272
	.01	4.433	4.397	4.364	4.307	4.259	4.218	4.079	3.948	3.825	3.707
49	.1	3.475	3.456	3.438	3.407	3.380	3.358	3.281	3.208	3.137	3.069
	.05	3.773	3.749	3.727	3.689	3.657	3.629	3.535	3.446	3.360	3.278
	.01	4.442	4.405	4.372	4.314	4.266	4.225	4.086	3.955	3.830	3.713
50	.1	3.484	3.464	3.446	3.415	3.388	3.366	3.289	3.214	3.143	3.075
	.05	3.782	3.757	3.735	3.697	3.664	3.637	3.542	3.452	3.366	3.283
	.01	4.450	4.413	4.380	4.322	4.274	4.232	4.093	3.961	3.836	3.718
51	.1	3.492	3.472	3.454	3.423	3.396	3.373	3.296	3.221	3.150	3.081
	.05	3.790	3.766	3.743	3.705	3.672	3.644	3.549	3.459	3.372	3.289
	.01	4.458	4.421	4.388	4.330	4.281	4.239	4.099	3.967	3.841	3.723
52	.1	3.500	3.480	3.462	3.430	3.404	3.381	3.303	3.228	3.156	3.087
	.05	3.798	3.773	3.751	3.712	3.679	3.651	3.556	3.465	3.378	3.295
	.01	4.466	4.429	4.395	4.337	4.288	4.246	4.106	3.972	3.847	3.728
53	.1	3.508	3.488	3.470	3.438	3.411	3.388	3.309	3.234	3.162	3.092
	.05	3.806	3.781	3.759	3.720	3.687	3.658	3.562	3.471	3.383	3.300
	.01	4.474	4.437	4.403	4.344	4.295	4.253	4.112	3.978	3.852	3.732
54	.1	3.516	3.496	3.477	3.445	3.418	3.395	3.316	3.240	3.168	3.098
	.05	3.814	3.789	3.766	3.727	3.694	3.665	3.569	3.477	3.389	3.305
	.01	4.481	4.444	4.410	4.351	4.302	4.260	4.118	3.984	3.857	3.737
55	.1	3.523	3.503	3.485	3.452	3.425	3.402	3.323	3.246	3.173	3.103
	.05	3.821	3.796	3.773	3.734	3.701	3.672	3.575	3.483	3.395	3.310
	.01	4.489	4.451	4.417	4.358	4.309	4.267	4.124	3.989	3.862	3.742
56	.1	3.531	3.510	3.492	3.460	3.432	3.409	3.329	3.253	3.179	3.109
	.05	3.828	3.803	3.781	3.741	3.707	3.679	3.582	3.489	3.400	3.315
	.01	4.496	4.459	4.425	4.365	4.315	4.273	4.130	3.995	3.867	3.746
57	.1	3.538	3.518	3.499	3.466	3.439	3.415	3.335	3.258	3.185	3.114
	.05	3.836	3.811	3.788	3.748	3.714	3.685	3.588	3.494	3.405	3.320
	.01	4.504	4.466	4.431	4.372	4.322	4.279	4.136	4.000	3.872	3.751
58	.1	3.545	3.525	3.506	3.473	3.446	3.422	3.342	3.264	3.190	3.119
	.05	3.843	3.818	3.795	3.754	3.721	3.692	3.594	3.500	3.411	3.325
	.01	4.511	4.473	4.438	4.379	4.328	4.286	4.141	4.005	3.877	3.755
59	.1	3.553	3.532	3.513	3.480	3.452	3.429	3.348	3.270	3.195	3.124
	.05	3.850	3.824	3.801	3.761	3.727	3.698	3.600	3.505	3.416	3.330
	.01	4.518	4.480	4.445	4.385	4.335	4.292	4.147	4.010	3.881	3.759
60	.1	3.560	3.539	3.520	3.487	3.459	3.435	3.354	3.276	3.201	3.129
	.05	3.857	3.831	3.808	3.768	3.733	3.704	3.605	3.511	3.421	3.335
	.01	4.525	4.486	4.452	4.391	4.341	4.298	4.153	4.015	3.886	3.764

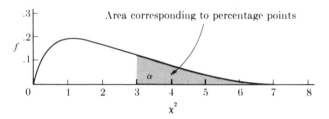

Area corresponding to percentage points

CL for Variance

$\chi^2 = $ chi sq.

$$P\left[\frac{(n-1)s_x^2}{\chi^2_{(0.025)(n-1)}} \leq \sigma_x^2 \leq \frac{(n-1)s_x^2}{\chi^2_{(.975)(n-1)}}\right] = 0.95 \text{ CL}$$

CL for Mean

t t-test
Table B

$$\bar{x} \pm t_{0.05(v)} \cdot s_{\bar{x}}$$

$$CL \quad P\left[\frac{(n-1) s_x^2}{\chi^2_{0.025(n-1)}} \leq \sigma_x^2 \leq \frac{(n-1) s_x^2}{\chi^2_{0.975(n-1)}}\right] = 0.95$$

Variance

TABLE **D**   Critical values of the chi-square distribution

This table furnishes critical values for the chi-square distribution for degrees of freedom $v = 1$ to 100 in increments of one. The percentage points (corresponding to $\alpha = 0.995, 0.975, 0.9, 0.5, 0.1, 0.05, 0.025, 0.01, 0.005$, and 0.001) represent the area to the right of the critical value of χ^2 in one tail of the distribution, as shown in the accompanying figure. The critical values of χ^2 are given to three significant decimal places, except when $\chi^2 > 100$, in which case they are given to two significant decimal places.

To find the critical value of χ^2 for a given number of degrees of freedom, look up v in the left (argument) column of the table and read off the desired values of χ^2 in that row. For example, for 8 degrees of freedom, $\chi^2_{.05[8]} = 15.507$ and $\chi^2_{.01[8]} = 20.090$. The last value indicates that 1% of the area of the chi-square distribution (for 8 degrees of freedom) is to the right of the value of $\chi^2 = 20.090$. For values of $v > 100$, compute approximate critical values of χ^2 using the following formula: $\chi^2_{\alpha[v]} = \frac{1}{2}(t_{2\alpha[\infty]} + \sqrt{2v - 1})^2$, where $t_{2\alpha[\infty]}$ comes from Table **B**. Thus $\chi^2_{.05[120]}$ is computed as $\frac{1}{2}(t_{.10[\infty]} + \sqrt{240 - 1})^2 = \frac{1}{2}(1.645 + \sqrt{239})^2 = \frac{1}{2}(17.10462)^2 = 146.284$. For $\alpha > 0.5$, employ $-t_{2(1-\alpha)[\infty]}$ in this formula. When $\alpha = 0.5$, $t_{2\alpha} = 0$.

There are numerous applications of the chi-square distribution (Section 7.6) in statistics. It is used to set confidence limits to sample variances (Section 7.7), to test the homogeneity of variances (Section 13.3) and of correlation coefficients (Section 15.5), and to test goodness of fit (Chapter 17).

Except for $\alpha = 0.001$, the values of chi-square from 1 to 30 degrees of freedom have been taken from a more extensive table by C. M. Thompson (*Biometrika* **32**:188–189, 1941) with permission of the publisher. Values between 31 and 100 degrees of freedom were approximated using the Cornish–Fisher asymptotic expansion following the account of M. Zelen and N. C. Severo, section 26.2.49 of M. Abramovitz and I. A. Stegun (eds.), *Handbook of Mathematical Functions* (U.S. National Bureau of Standards, 1964). The values of the Hermite polynomials were computed, instead of using the tables in the source above. All values for $\alpha = 0.001$ were taken from H. L. Harter, *New Tables of the Incomplete Gamma-Function Ratio and of Percentage Points of the Chi-Square and Beta Distribution* (U.S. Government Printing Office, 1964).

TABLE D Critical values of the chi-square distribution

$\nu \backslash \alpha$	.995	.975	.9	.5	.1	.05	.025	.01	.005	.001	α / ν
1	0.000	0.000	0.016	0.455	2.706	3.841	5.024	6.635	7.879	10.828	1
2	0.010	0.051	0.211	1.386	4.605	5.991	7.378	9.210	10.597	13.816	2
3	0.072	0.216	0.584	2.366	6.251	7.815	9.348	11.345	12.838	16.266	3
4	0.207	0.484	1.064	3.357	7.779	9.488	11.143	13.277	14.860	18.467	4
5	0.412	0.831	1.610	4.351	9.236	11.070	12.832	15.086	16.750	20.515	5
6	0.676	1.237	2.204	5.348	10.645	12.592	14.449	16.812	18.548	22.458	6
7	0.989	1.690	2.833	6.346	12.017	14.067	16.013	18.475	20.278	24.322	7
8	1.344	2.180	3.490	7.344	13.362	15.507	17.535	20.090	21.955	26.124	8
9	1.735	2.700	4.168	8.343	14.684	16.919	19.023	21.666	23.589	27.877	9
10	2.156	3.247	4.865	9.342	15.987	18.307	20.483	23.209	25.188	29.588	10
11	2.603	3.816	5.578	10.341	17.275	19.675	21.920	24.725	26.757	31.264	11
12	3.074	4.404	6.304	11.340	18.549	21.026	23.337	26.217	28.300	32.910	12
13	3.565	5.009	7.042	12.340	19.812	22.362	24.736	27.688	29.819	34.528	13
14	4.075	5.629	7.790	13.339	21.064	23.685	26.119	29.141	31.319	36.123	14
15	4.601	6.262	8.547	14.339	22.307	24.996	27.488	30.578	32.801	37.697	15
16	5.142	6.908	9.312	15.338	23.542	26.296	28.845	32.000	34.267	39.252	16
17	5.697	7.564	10.085	16.338	24.769	27.587	30.191	33.409	35.718	40.790	17
18	6.265	8.231	10.865	17.338	25.989	28.869	31.526	34.805	37.156	42.312	18
19	6.844	8.907	11.651	18.338	27.204	30.144	32.852	36.191	38.582	43.820	19
20	7.434	9.591	12.443	19.337	28.412	31.410	34.170	37.566	39.997	45.315	20
21	8.034	10.283	13.240	20.337	29.615	32.670	35.479	38.932	41.401	46.797	21
22	8.643	10.982	14.042	21.337	30.813	33.924	36.781	40.289	42.796	48.268	22
23	9.260	11.688	14.848	22.337	32.007	35.172	38.076	41.638	44.181	49.728	23
24	9.886	12.401	15.659	23.337	33.196	36.415	39.364	42.980	45.558	51.179	24
25	10.520	13.120	16.473	24.337	34.382	37.652	40.646	44.314	46.928	52.620	25
26	11.160	13.844	17.292	25.336	35.563	38.885	41.923	45.642	48.290	54.052	26
27	11.808	14.573	18.114	26.336	36.741	40.113	43.194	46.963	49.645	55.476	27
28	12.461	15.308	18.939	27.336	37.916	41.337	44.461	48.278	50.993	56.892	28
29	13.121	16.047	19.768	28.336	39.088	42.557	45.722	49.588	52.336	58.301	29
30	13.787	16.791	20.599	29.336	40.256	43.773	46.979	50.892	53.672	59.703	30
31	14.458	17.539	21.434	30.336	41.422	44.985	48.232	52.191	55.003	61.098	31
32	15.134	18.291	22.271	31.336	42.585	46.194	49.480	53.486	56.329	62.487	32
33	15.815	19.047	23.110	32.336	43.745	47.400	50.725	54.776	57.649	63.870	33
34	16.501	19.806	23.952	33.336	44.903	48.602	51.966	56.061	58.964	65.247	34
35	17.192	20.569	24.797	34.336	46.059	49.802	53.203	57.342	60.275	66.619	35
36	17.887	21.336	25.643	35.336	47.212	50.998	54.437	58.619	61.582	67.985	36
37	18.586	22.106	26.492	36.335	48.363	52.192	55.668	59.892	62.884	69.346	37
38	19.289	22.878	27.343	37.335	49.513	53.384	56.896	61.162	64.182	70.703	38
39	19.996	23.654	28.196	38.335	50.660	54.572	58.120	62.428	65.476	72.055	39
40	20.707	24.433	29.051	39.335	51.805	55.758	59.342	63.691	66.766	73.402	40
41	21.421	25.215	29.907	40.335	52.949	56.942	60.561	64.950	68.053	74.745	41
42	22.138	25.999	30.765	41.335	54.090	58.124	61.777	66.206	69.336	76.084	42
43	22.859	26.785	31.625	42.335	55.230	59.304	62.990	67.459	70.616	77.419	43
44	23.584	27.575	32.487	43.335	56.369	60.481	64.202	68.710	71.893	78.750	44
45	24.311	28.366	33.350	44.335	57.505	61.656	65.410	69.957	73.166	80.077	45
46	25.042	29.160	34.215	45.335	58.641	62.830	66.617	71.201	74.437	81.400	46
47	25.775	29.956	35.081	46.335	59.774	64.001	67.821	72.443	75.704	82.720	47
48	26.511	30.755	35.949	47.335	60.907	65.171	69.023	73.683	76.969	84.037	48
49	27.249	31.555	36.818	48.335	62.038	66.339	70.222	74.919	78.231	85.351	49
50	27.991	32.357	37.689	49.335	63.167	67.505	71.420	76.154	79.490	86.661	50

TABLE D Critical values of the chi-square distribution (*continued*)

$\nu \backslash \alpha$	.995	.975	.9	.5	.1	.05	.025	.01	.005	.001	α / ν
51	28.735	33.162	38.560	50.335	64.295	68.669	72.616	77.386	80.747	87.968	51
52	29.481	33.968	39.433	51.335	65.422	69.832	73.810	78.616	82.001	89.272	52
53	30.230	34.776	40.308	52.335	66.548	70.993	75.002	79.843	83.253	90.573	53
54	30.981	35.586	41.183	53.335	67.673	72.153	76.192	81.069	84.502	91.872	54
55	31.735	36.398	42.060	54.335	68.796	73.311	77.380	82.292	85.749	93.168	55
56	32.490	37.212	42.937	55.335	69.918	74.468	78.567	83.513	86.994	94.460	56
57	33.248	38.027	43.816	56.335	71.040	75.624	79.752	84.733	88.237	95.751	57
58	34.008	38.844	44.696	57.335	72.160	76.778	80.936	85.950	89.477	97.039	58
59	34.770	39.662	45.577	58.335	73.279	77.931	82.117	87.166	90.715	98.324	59
60	35.534	40.482	46.459	59.335	74.397	79.082	83.298	88.379	91.952	99.607	60
61	36.300	41.303	47.342	60.335	75.514	80.232	84.476	89.591	93.186	100.888	61
62	37.068	42.126	48.226	61.335	76.630	81.381	85.654	90.802	94.419	102.166	62
63	37.838	42.950	49.111	62.335	77.745	82.529	86.830	92.010	95.649	103.442	63
64	38.610	43.776	49.996	63.335	78.860	83.675	88.004	93.217	96.878	104.716	64
65	39.383	44.603	50.883	64.335	79.973	84.821	89.177	94.422	98.105	105.988	65
66	40.158	45.431	51.770	65.335	81.085	85.965	90.349	95.626	99.331	107.258	66
67	40.935	46.261	52.659	66.335	82.197	87.108	91.519	96.828	100.55	108.526	67
68	41.713	47.092	53.548	67.334	83.308	88.250	92.689	98.028	101.78	109.791	68
69	42.494	47.924	54.438	68.334	84.418	89.391	93.856	99.228	103.00	111.055	69
70	43.275	48.758	55.329	69.334	85.527	90.531	95.023	100.43	104.21	112.317	70
71	44.058	49.592	56.221	70.334	86.635	91.670	96.189	101.62	105.43	113.577	71
72	44.843	50.428	57.113	71.334	87.743	92.808	97.353	102.82	106.65	114.835	72
73	45.629	51.265	58.006	72.334	88.850	93.945	98.516	104.01	107.86	116.092	73
74	46.417	52.103	58.900	73.334	89.956	95.081	99.678	105.20	109.07	117.346	74
75	47.206	52.942	59.795	74.334	91.061	96.217	100.84	106.39	110.29	118.599	75
76	47.997	53.782	60.690	75.334	92.166	97.351	102.00	107.58	111.50	119.850	76
77	48.788	54.623	61.586	76.334	93.270	98.484	103.16	108.77	112.70	121.100	77
78	49.582	55.466	62.483	77.334	94.373	99.617	104.32	109.96	113.91	122.348	78
79	50.376	56.309	63.380	78.334	95.476	100.75	105.47	111.14	115.12	123.594	79
80	51.172	57.153	64.278	79.334	96.578	101.88	106.63	112.33	116.32	124.839	80
81	51.969	57.998	65.176	80.334	97.680	103.01	107.78	113.51	117.52	126.082	81
82	52.767	58.845	66.076	81.334	98.780	104.14	108.94	114.69	118.73	127.324	82
83	53.567	59.692	66.976	82.334	99.880	105.27	110.09	115.88	119.93	128.565	83
84	54.368	60.540	67.876	83.334	100.98	106.39	111.24	117.06	121.13	129.804	84
85	55.170	61.389	68.777	84.334	102.08	107.52	112.39	118.24	122.32	131.041	85
86	55.973	62.239	69.679	85.334	103.18	108.65	113.54	119.41	123.52	132.277	86
87	56.777	63.089	70.581	86.334	104.28	109.77	114.69	120.59	124.72	133.512	87
88	57.582	63.941	71.484	87.334	105.37	110.90	115.84	121.77	125.91	134.745	88
89	58.389	64.793	72.387	88.334	106.47	112.02	116.99	122.94	127.11	135.978	89
90	59.196	65.647	73.291	89.334	107.56	113.15	118.14	124.12	128.30	137.208	90
91	60.005	66.501	74.196	90.334	108.66	114.27	119.28	125.29	129.49	138.438	91
92	60.815	67.356	75.101	91.334	109.76	115.39	120.43	126.46	130.68	139.666	92
93	61.625	68.211	76.006	92.334	110.85	116.51	121.57	127.63	131.87	140.893	93
94	62.437	69.068	76.912	93.334	111.94	117.63	122.72	128.80	133.06	142.119	94
95	63.250	69.925	77.818	94.334	113.04	118.75	123.86	129.97	134.25	143.344	95
96	64.063	70.783	78.725	95.334	114.13	119.87	125.00	131.14	135.43	144.567	96
97	64.878	71.642	79.633	96.334	115.22	120.99	126.14	132.31	136.62	145.789	97
98	65.694	72.501	80.541	97.334	116.32	122.11	127.28	133.48	137.80	147.010	98
99	66.510	73.361	81.449	98.334	117.41	123.23	128.42	134.64	138.99	148.230	99
100	67.328	74.222	82.358	99.334	118.50	124.34	129.56	135.81	140.17	149.449	100

TABLE **E**   Critical values of the chi-square distribution based on Šidák's multiplicative inequality

This table furnishes critical values of the chi-square distribution for unusual percentage points not found in standard chi-square tables such as Table **D.** These unusual percentage points α' are obtained from Šidák's multiplicative inequality: $\alpha' = 1 - (1 - \alpha)^{1/k}$, where α is the experimentwise error rate (also known as the familywise type I error rate) and k is the number of comparisons intended. Values of $\chi^2_{\alpha'}$, are furnished for degrees of freedom v from 1 to 20 in increments of one. Arguments for k, the number of comparisons, are given from 1 to 60 in increments of one. Two levels of experimentwise α (0.05 and 0.01) are given for each combination of v and k in the table.

To find the critical value for a 5% experimentwise error based on a chi-square with $v = 6$ degrees of freedom and making $k = 5$ comparisons overall, we enter the table at $k = 5$, $v = 6$, and $\alpha = .05$ to obtain $\chi^2_{.010206[6]} = 16.760$. This table is one-tailed.

This table is employed for unplanned multiple-comparisons tests for the homogeneity of replicates in replicated tests of goodness of fit (Section 17.3).

The table was computed directly using a modification of the Hewlett Packard 9830 Statistical Distributions Pack, volume 1. All computations were carried out to 12 decimal places.

TABLE **E** Critical values of the chi-square distribution based on
Šidák's multiplicative inequality

Degrees of freedom v

k	α	1	2	3	4	5	6	7	8	9	10
1	.05	3.841	5.991	7.815	9.488	11.070	12.592	14.067	15.507	16.919	18.307
	.01	6.635	9.210	11.345	13.277	15.086	16.812	18.475	20.090	21.666	23.209
2	.05	5.002	7.352	9.320	11.113	12.801	14.416	15.978	17.498	18.985	20.444
	.01	7.875	10.592	12.833	14.855	16.744	18.541	20.271	21.948	23.582	25.181
3	.05	5.701	8.155	10.198	12.054	13.797	15.462	17.070	18.633	20.160	21.657
	.01	8.609	11.401	13.699	15.770	17.702	19.539	21.305	23.015	24.681	26.310
4	.05	6.205	8.726	10.820	12.718	14.497	16.196	17.834	19.426	20.980	22.502
	.01	9.134	11.975	14.312	16.415	18.377	20.240	22.031	23.765	25.452	27.102
5	.05	6.599	9.169	11.301	13.230	15.037	16.760	18.422	20.035	21.608	23.150
	.01	9.542	12.421	14.787	16.915	18.898	20.781	22.591	24.342	26.046	27.711
6	.05	6.922	9.532	11.693	13.647	15.476	17.219	18.898	20.528	22.118	23.675
	.01	9.877	12.785	15.174	17.322	19.322	21.222	23.046	24.811	26.528	28.205
7	.05	7.197	9.840	12.024	13.998	15.845	17.604	19.299	20.943	22.546	24.115
	.01	10.161	13.094	15.501	17.665	19.680	21.593	23.429	25.205	26.933	28.621
8	.05	7.437	10.106	12.311	14.302	16.164	17.937	19.644	21.300	22.915	24.494
	.01	10.407	13.360	15.784	17.962	19.989	21.913	23.760	25.546	27.283	28.980
9	.05	7.648	10.341	12.563	14.569	16.445	18.230	19.948	21.614	23.238	24.827
	.01	10.624	13.596	16.034	18.223	20.262	22.195	24.051	25.846	27.591	29.295
10	.05	7.838	10.551	12.789	14.808	16.695	18.491	20.219	21.894	23.526	25.124
	.01	10.819	13.806	16.257	18.457	20.505	22.447	24.311	26.113	27.865	29.576
11	.05	8.010	10.741	12.993	15.024	16.921	18.726	20.463	22.146	23.786	25.391
	.01	10.996	13.997	16.458	18.668	20.724	22.674	24.545	26.354	28.113	29.830
12	.05	8.167	10.914	13.179	15.220	17.127	18.940	20.685	22.376	24.023	25.634
	.01	11.157	14.171	16.642	18.860	20.924	22.881	24.759	26.574	28.339	30.061
13	.05	8.312	11.074	13.350	15.401	17.316	19.137	20.889	22.586	24.240	25.857
	.01	11.305	14.331	16.811	19.037	21.108	23.072	24.955	26.776	28.546	30.273
14	.05	8.447	11.222	13.508	15.568	17.491	19.319	21.078	22.781	24.440	26.062
	.01	11.443	14.479	16.968	19.201	21.278	23.248	25.137	26.962	28.737	30.469
15	.05	8.572	11.360	13.655	15.723	17.653	19.488	21.253	22.962	24.626	26.254
	.01	11.571	14.617	17.113	19.353	21.436	23.411	25.305	27.136	28.915	30.651
16	.05	8.689	11.489	13.793	15.869	17.805	19.646	21.416	23.131	24.800	26.432
	.01	11.691	14.746	17.250	19.495	21.584	23.564	25.463	27.298	29.081	30.821
17	.05	8.800	11.610	13.922	16.005	17.948	19.794	21.569	23.289	24.963	26.599
	.01	11.804	14.867	17.377	19.629	21.723	23.708	25.611	27.450	29.237	30.981
18	.05	8.904	11.724	14.044	16.133	18.082	19.934	21.714	23.438	25.116	26.756
	.01	11.910	14.982	17.498	19.755	21.854	23.843	25.750	27.593	29.384	31.131
19	.05	9.003	11.832	14.159	16.254	18.209	20.066	21.850	23.578	25.261	26.905
	.01	12.011	15.090	17.612	19.874	21.977	23.970	25.881	27.728	29.522	31.273
20	.05	9.096	11.934	14.269	16.370	18.329	20.190	21.979	23.711	25.398	27.046
	.01	12.107	15.192	17.720	19.987	22.094	24.092	26.006	27.856	29.654	31.407

TABLE E Critical values of the chi-square distribution based on
Šidák's multiplicative inequality (*continued*)

Degrees of freedom v

k	α	1	2	3	4	5	6	7	8	9	10
21	.05	9.185	12.032	14.373	16.479	18.443	20.309	22.102	23.838	25.528	27.179
	.01	12.198	15.290	17.823	20.094	22.206	24.207	26.125	27.978	29.778	31.535
22	.05	9.270	12.125	14.472	16.583	18.552	20.422	22.219	23.959	25.652	27.306
	.01	12.285	15.383	17.921	20.196	22.312	24.316	26.238	28.094	29.897	31.657
23	.05	9.352	12.214	14.566	16.682	18.656	20.530	22.330	24.074	25.770	27.428
	.01	12.367	15.472	18.014	20.294	22.413	24.421	26.345	28.205	30.011	31.773
24	.05	9.430	12.299	14.657	16.778	18.755	20.633	22.437	24.183	25.883	27.544
	.01	12.447	15.557	18.104	20.388	22.510	24.521	26.448	28.310	30.119	31.884
25	.05	9.505	12.380	14.743	16.869	18.850	20.732	22.539	24.289	25.991	27.655
	.01	12.523	15.638	18.190	20.477	22.603	24.617	26.547	28.412	30.223	31.990
26	.05	9.576	12.459	14.827	16.957	18.942	20.827	22.637	24.390	26.095	27.762
	.01	12.596	15.717	18.272	20.563	22.692	24.709	26.642	28.509	30.323	32.092
27	.05	9.646	12.534	14.907	17.041	19.030	20.918	22.732	24.487	26.195	27.864
	.01	12.667	15.792	18.352	20.646	22.778	24.798	26.733	28.603	30.419	32.191
28	.05	9.712	12.607	14.984	17.122	19.114	21.006	22.823	24.581	26.291	27.963
	.01	12.735	15.865	18.428	20.726	22.861	24.883	26.821	28.693	30.512	32.285
29	.05	9.777	12.677	15.059	17.200	19.196	21.091	22.910	24.671	26.384	28.058
	.01	12.801	15.935	18.502	20.803	22.941	24.966	26.906	28.780	30.601	32.376
30	.05	9.839	12.744	15.131	17.276	19.275	21.172	22.995	24.758	26.474	28.150
	.01	12.864	16.003	18.573	20.877	23.018	25.045	26.988	28.864	30.687	32.464
31	.05	9.899	12.810	15.200	17.349	19.351	21.251	23.076	24.842	26.560	28.238
	.01	12.925	16.069	18.642	20.949	23.092	25.122	27.067	28.946	30.770	32.550
32	.05	9.958	12.873	15.268	17.420	19.425	21.328	23.155	24.924	26.644	28.324
	.01	12.985	16.132	18.709	21.018	23.164	25.196	27.144	29.024	30.851	32.632
33	.05	10.014	12.935	15.333	17.488	19.496	21.402	23.232	25.002	26.725	28.407
	.01	13.042	16.194	18.774	21.086	23.234	25.268	27.218	29.100	30.929	32.712
34	.05	10.069	12.995	15.396	17.555	19.565	21.474	23.306	25.079	26.803	28.488
	.01	13.098	16.253	18.836	21.151	23.302	25.338	27.290	29.174	31.004	32.789
35	.05	10.123	13.053	15.458	17.619	19.633	21.543	23.378	25.153	26.879	28.566
	.01	13.153	16.311	18.897	21.215	23.367	25.406	27.359	29.246	31.077	32.864
36	.05	10.175	13.109	15.517	17.682	19.698	21.611	23.448	25.225	26.953	28.642
	.01	13.205	16.368	18.956	21.276	23.431	25.472	27.427	29.315	31.149	32.937
37	.05	10.225	13.164	15.576	17.743	19.761	21.677	23.516	25.295	27.025	28.715
	.01	13.257	16.422	19.014	21.336	23.493	25.536	27.493	29.383	31.218	33.008
38	.05	10.274	13.217	15.632	17.802	19.823	21.741	23.582	25.363	27.095	28.787
	.01	13.307	16.476	19.070	21.394	23.554	25.598	27.557	29.449	31.285	33.076
39	.05	10.322	13.269	15.687	17.860	19.883	21.803	23.646	25.429	27.163	28.857
	.01	13.355	16.528	19.124	21.451	23.613	25.659	27.620	29.513	31.351	33.143
40	.05	10.369	13.319	15.741	17.916	19.942	21.864	23.709	25.494	27.230	28.925
	.01	13.403	16.578	19.178	21.507	23.670	25.718	27.680	29.575	31.415	33.209

TABLE **E** Critical values of the chi-square distribution based on Šidák's multiplicative inequality (*continued*)

Degrees of freedom v

k	α	1	2	3	4	5	6	7	8	9	10
41	.05	10.415	13.369	15.793	17.971	19.999	21.923	23.770	25.557	27.294	28.991
	.01	13.449	16.628	19.229	21.561	23.726	25.776	27.740	29.636	31.477	33.273
42	.05	10.459	13.417	15.844	18.025	20.055	21.981	23.830	25.618	27.357	29.056
	.01	13.494	16.676	19.280	21.613	23.780	25.832	27.798	29.695	31.538	33.335
43	.05	10.502	13.464	15.894	18.077	20.109	22.037	23.888	25.678	27.419	29.119
	.01	13.539	16.723	19.329	21.665	23.834	25.887	27.854	29.753	31.597	33.395
44	.05	10.545	13.510	15.943	18.128	20.162	22.092	23.945	25.736	27.479	29.180
	.01	13.582	16.769	19.378	21.715	23.885	25.941	27.909	29.810	31.655	33.454
45	.05	10.586	13.555	15.990	18.178	20.214	22.146	24.000	25.794	27.537	29.240
	.01	13.624	16.814	19.425	21.764	23.936	25.993	27.963	29.865	31.712	33.512
46	.05	10.627	13.599	16.037	18.226	20.265	22.199	24.055	25.849	27.595	29.299
	.01	13.665	16.858	19.471	21.812	23.986	26.044	28.016	29.919	31.767	33.569
47	.05	10.667	13.642	16.082	18.274	20.315	22.250	24.108	25.904	27.651	29.356
	.01	13.706	16.901	19.516	21.859	24.035	26.094	28.067	29.972	31.821	33.624
48	.05	10.706	13.684	16.127	18.321	20.363	22.300	24.160	25.958	27.706	29.413
	.01	13.745	16.943	19.560	21.905	24.082	26.143	28.118	30.024	31.874	33.678
49	.05	10.744	13.725	16.170	18.367	20.411	22.350	24.210	26.010	27.759	29.468
	.01	13.784	16.984	19.603	21.950	24.129	26.191	28.167	30.074	31.926	33.731
50	.05	10.781	13.765	16.213	18.411	20.457	22.398	24.260	26.061	27.812	29.522
	.01	13.822	17.025	19.646	21.994	24.174	26.238	28.215	30.124	31.977	33.783
51	.05	10.818	13.805	16.255	18.455	20.503	22.445	24.309	26.111	27.864	29.574
	.01	13.859	17.064	19.687	22.037	24.219	26.284	28.263	30.172	32.026	33.834
52	.05	10.854	13.844	16.296	18.498	20.548	22.492	24.357	26.160	27.914	29.626
	.01	13.895	17.103	19.728	22.079	24.263	26.330	28.309	30.220	32.075	33.884
53	.05	10.889	13.882	16.337	18.540	20.592	22.537	24.404	26.209	27.964	29.677
	.01	13.931	17.141	19.768	22.121	24.306	26.374	28.355	30.267	32.123	33.933
54	.05	10.924	13.919	16.376	18.582	20.635	22.582	24.450	26.256	28.012	29.727
	.01	13.966	17.178	19.807	22.162	24.348	26.417	28.399	30.313	32.170	33.981
55	.05	10.958	13.956	16.415	18.622	20.677	22.625	24.495	26.302	28.060	29.775
	.01	14.001	17.215	19.846	22.202	24.389	26.460	28.443	30.358	32.216	34.028
56	.05	10.991	13.992	16.453	18.662	20.718	22.668	24.539	26.348	28.107	29.823
	.01	14.035	17.251	19.883	22.241	24.430	26.502	28.486	30.402	32.261	34.074
57	.05	11.024	14.027	16.490	18.702	20.759	22.710	24.583	26.393	28.153	29.870
	.01	14.068	17.287	19.920	22.280	24.470	26.543	28.529	30.445	32.306	34.119
58	.05	11.056	14.062	16.527	18.740	20.799	22.752	24.625	26.437	28.198	29.917
	.01	14.101	17.321	19.957	22.317	24.509	26.584	28.570	30.488	32.349	34.164
59	.05	11.088	14.096	16.563	18.778	20.838	22.793	24.667	26.480	28.242	29.962
	.01	14.133	17.355	19.993	22.355	24.548	26.623	28.611	30.530	32.392	34.207
60	.05	11.119	14.130	16.599	18.815	20.877	22.833	24.709	26.522	28.285	30.007
	.01	14.165	17.389	20.028	22.391	24.586	26.662	28.651	30.571	32.434	34.251

TABLE E Critical values of the chi-square distribution based on Šidák's multiplicative inequality (*continued*)

Degrees of freedom v

k	α	11	12	13	14	15	16	17	18	19	20
1	.05	19.675	21.026	22.362	23.685	24.996	26.296	27.587	28.869	30.144	31.410
	.01	24.725	26.217	27.688	29.141	30.578	32.000	33.409	34.805	36.191	37.566
2	.05	21.880	23.295	24.693	26.075	27.444	28.800	30.145	31.479	32.804	34.121
	.01	26.750	28.292	29.812	31.312	32.793	34.259	35.710	37.148	38.574	39.988
3	.05	23.128	24.578	26.009	27.423	28.822	30.208	31.582	32.944	34.296	35.639
	.01	27.908	29.478	31.024	32.549	34.056	35.545	37.020	38.480	39.928	41.364
4	.05	23.998	25.471	26.924	28.360	29.780	31.185	32.578	33.959	35.330	36.691
	.01	28.719	30.308	31.872	33.415	34.938	36.444	37.934	39.410	40.873	42.323
5	.05	24.664	26.155	27.624	29.076	30.511	31.932	33.339	34.735	36.119	37.493
	.01	29.342	30.946	32.524	34.079	35.615	37.134	38.636	40.123	41.597	43.059
6	.05	25.203	26.707	28.190	29.655	31.102	32.535	33.954	35.360	36.755	38.140
	.01	29.849	31.463	33.052	34.618	36.165	37.693	39.204	40.701	42.184	43.655
7	.05	25.655	27.171	28.665	30.140	31.597	33.040	34.468	35.884	37.288	38.682
	.01	30.274	31.898	33.496	35.071	36.626	38.162	39.682	41.186	42.677	44.155
8	.05	26.045	27.570	29.073	30.557	32.023	33.474	34.911	36.334	37.746	39.148
	.01	30.641	32.273	33.879	35.462	37.024	38.567	40.093	41.604	43.101	44.586
9	.05	26.386	27.920	29.431	30.923	32.397	33.855	35.298	36.729	38.148	39.555
	.01	30.964	32.603	34.215	35.805	37.373	38.922	40.455	41.971	43.474	44.964
10	.05	26.691	28.232	29.750	31.248	32.729	34.193	35.643	37.080	38.504	39.918
	.01	31.252	32.897	34.515	36.110	37.684	39.239	40.776	42.298	43.806	45.300
11	.05	26.965	28.512	30.037	31.542	33.028	34.498	35.954	37.396	38.826	40.244
	.01	31.511	33.162	34.786	36.386	37.965	39.524	41.067	42.593	44.105	45.604
12	.05	27.214	28.768	30.298	31.808	33.300	34.775	36.236	37.683	39.117	40.540
	.01	31.748	33.403	35.032	36.637	38.220	39.784	41.331	42.861	44.377	45.880
13	.05	27.442	29.002	30.537	32.052	33.549	35.029	36.494	37.946	39.385	40.812
	.01	31.964	33.625	35.258	36.867	38.454	40.022	41.573	43.107	44.627	46.133
14	.05	27.653	29.218	30.758	32.278	33.779	35.263	36.733	38.188	39.631	41.062
	.01	32.165	33.829	35.466	37.079	38.671	40.242	41.796	43.334	44.857	46.367
15	.05	27.849	29.418	30.963	32.487	33.992	35.481	36.954	38.413	39.860	41.295
	.01	32.351	34.019	35.660	37.277	38.871	40.447	42.004	43.545	45.071	46.584
16	.05	28.032	29.605	31.154	32.682	34.191	35.684	37.160	38.623	40.073	41.511
	.01	32.525	34.197	35.841	37.461	39.059	40.637	42.198	43.742	45.271	46.787
17	.05	28.204	29.781	31.334	32.865	34.378	35.874	37.354	38.820	40.273	41.714
	.01	32.688	34.363	36.011	37.634	39.235	40.816	42.380	43.926	45.458	46.976
18	.05	28.365	29.946	31.502	33.037	34.553	36.052	37.535	39.005	40.461	41.905
	.01	32.842	34.520	36.171	37.797	39.401	40.985	42.551	44.100	45.635	47.155
19	.05	28.517	30.101	31.661	33.200	34.719	36.221	37.707	39.179	40.638	42.085
	.01	32.987	34.668	36.322	37.950	39.557	41.144	42.712	44.264	45.801	47.324
20	.05	28.661	30.249	31.812	33.353	34.876	36.380	37.869	39.344	40.806	42.255
	.01	33.124	34.808	36.465	38.096	39.705	41.294	42.865	44.419	45.958	47.484

TABLE E Critical values of the chi-square distribution based on Šidák's multiplicative inequality (*continued*)

Degrees of freedom ν

k	α	11	12	13	14	15	16	17	18	19	20
21	.05	28.798	30.389	31.955	33.499	35.024	36.532	38.023	39.501	40.965	42.417
	.01	33.254	34.941	36.600	38.234	39.846	41.437	43.010	44.567	46.108	47.635
22	.05	28.928	30.522	32.091	33.638	35.166	36.676	38.170	39.650	41.116	42.571
	.01	33.379	35.068	36.730	38.366	39.980	41.573	43.149	44.707	46.251	47.780
23	.05	29.053	30.649	32.221	33.771	35.301	36.813	38.310	39.792	41.261	42.717
	.01	33.497	35.189	36.853	38.491	40.107	41.703	43.281	44.841	46.386	47.918
24	.05	29.171	30.771	32.345	33.897	35.430	36.945	38.444	39.928	41.399	42.857
	.01	33.611	35.305	36.971	38.612	40.230	41.827	43.407	44.969	46.516	48.049
25	.05	29.285	30.887	32.464	34.018	35.553	37.071	38.572	40.058	41.531	42.991
	.01	33.720	35.416	37.084	38.727	40.347	41.946	43.528	45.092	46.641	48.175
26	.05	29.395	30.999	32.578	34.135	35.672	37.191	38.694	40.183	41.658	43.120
	.01	33.824	35.523	37.192	38.837	40.459	42.061	43.644	45.210	46.760	48.297
27	.05	29.499	31.106	32.687	34.247	35.786	37.307	38.812	40.303	41.779	43.244
	.01	33.924	35.625	37.297	38.943	40.567	42.170	43.755	45.323	46.875	48.413
28	.05	29.600	31.209	32.793	34.354	35.895	37.419	38.926	40.418	41.897	43.363
	.01	34.021	35.723	37.397	39.045	40.671	42.276	43.862	45.432	46.986	48.525
29	.05	29.698	31.309	32.895	34.458	36.001	37.526	39.035	40.529	42.009	43.477
	.01	34.114	35.818	37.494	39.144	40.771	42.378	43.966	45.537	47.092	48.633
30	.05	29.792	31.405	32.993	34.558	36.103	37.630	39.141	40.636	42.118	43.588
	.01	34.204	35.910	37.587	39.239	40.868	42.476	44.066	45.638	47.195	48.737
31	.05	29.883	31.498	33.087	34.654	36.201	37.730	39.243	40.740	42.224	43.695
	.01	34.291	35.999	37.678	39.331	40.961	42.571	44.162	45.736	47.294	48.838
32	.05	29.970	31.588	33.179	34.748	36.297	37.827	39.341	40.840	42.325	43.798
	.01	34.375	36.084	37.765	39.420	41.052	42.663	44.255	45.831	47.390	48.935
33	.05	30.055	31.674	33.268	34.838	36.389	37.921	39.436	40.937	42.424	43.898
	.01	34.456	36.168	37.850	39.506	41.139	42.752	44.346	45.922	47.483'	49.029
34	.05	30.138	31.759	33.354	34.926	36.478	38.012	39.529	41.031	42.519	43.994
	.01	34.535	36.248	37.932	39.590	41.224	42.838	44.433	46.011	47.573	49.121
35	.05	30.218	31.840	33.437	35.011	36.565	38.100	39.619	41.122	42.612	44.088
	.01	34.612	36.326	38.011	39.671	41.307	42.922	44.518	46.097	47.661	49.209
36	.05	30.295	31.920	33.518	35.094	36.649	38.186	39.706	41.210	42.701	44.180
	.01	34.686	36.402	38.089	39.749	41.387	43.003	44.601	46.181	47.746	49.296
37	.05	30.371	31.997	33.597	35.174	36.731	38.269	39.790	41.296	42.789	44.268
	.01	34.758	36.476	38.164	39.826	41.464	43.082	44.681	46.263	47.828	49.379
38	.05	30.444	32.072	33.673	35.252	36.810	38.350	39.872	41.380	42.873	44.354
	.01	34.829	36.547	38.237	39.900	41.540	43.159	44.759	46.342	47.908	49.461
39	.05	30.516	32.145	33.748	35.328	36.887	38.428	39.952	41.461	42.956	44.438
	.01	34.897	36.617	38.308	39.972	41.614	43.234	44.835	46.419	47.987	49.540
40	.05	30.585	32.216	33.820	35.402	36.963	38.505	40.030	41.540	43.036	44.519
	.01	34.964	36.685	38.377	40.043	41.685	43.307	44.909	46.494	48.063	49.617

TABLE E Critical values of the chi-square distribution based on Šidák's multiplicative inequality (*continued*)

Degrees of freedom v

k	α	11	12	13	14	15	16	17	18	19	20
41	.05	30.653	32.285	33.891	35.474	37.036	38.580	40.106	41.618	43.115	44.599
	.01	35.029	36.752	38.445	40.112	41.755	43.378	44.981	46.567	48.137	49.692
42	.05	30.719	32.353	33.960	35.544	37.108	38.652	40.180	41.693	43.191	44.676
	.01	35.092	36.816	38.511	40.179	41.823	43.447	45.051	46.638	48.209	49.765
43	.05	30.784	32.419	34.027	35.613	37.177	38.723	40.252	41.766	43.266	44.752
	.01	35.154	36.879	38.575	40.244	41.890	43.514	45.120	46.708	48.280	49.837
44	.05	30.847	32.483	34.093	35.680	37.246	38.793	40.323	41.838	43.338	44.826
	.01	35.215	36.941	38.638	40.308	41.955	43.580	45.187	46.776	48.349	49.907
45	.05	30.908	32.546	34.157	35.745	37.312	38.860	40.392	41.908	43.409	44.898
	.01	35.274	37.001	38.699	40.370	42.018	43.645	45.252	46.842	48.416	49.975
46	.05	30.968	32.607	34.220	35.809	37.377	38.927	40.459	41.976	43.479	44.968
	.01	35.331	37.060	38.759	40.431	42.080	43.708	45.316	46.907	48.482	50.042
47	.05	31.027	32.667	34.281	35.871	37.441	38.991	40.525	42.043	43.547	45.037
	.01	35.388	37.118	38.817	40.491	42.141	43.769	45.379	46.970	48.546	50.107
48	.05	31.084	32.726	34.341	35.932	37.503	39.055	40.589	42.108	43.613	45.104
	.01	35.443	37.174	38.875	40.549	42.200	43.830	45.440	47.033	48.609	50.171
49	.05	31.141	32.783	34.400	35.992	37.564	39.117	40.652	42.172	43.678	45.170
	.01	35.497	37.229	38.931	40.606	42.258	43.889	45.500	47.093	48.671	50.233
50	.05	31.196	32.840	34.457	36.051	37.623	39.177	40.714	42.235	43.741	45.235
	.01	35.550	37.283	38.986	40.662	42.315	43.946	45.558	47.153	48.731	50.294
51	.05	31.250	32.895	34.513	36.108	37.682	39.237	40.774	42.296	43.804	45.298
	.01	35.602	37.336	39.040	40.717	42.371	44.003	45.616	47.211	48.790	50.354
52	.05	31.303	32.949	34.568	36.164	37.739	39.295	40.833	42.356	43.865	45.360
	.01	35.653	37.388	39.093	40.771	42.425	44.058	45.672	47.268	48.848	50.413
53	.05	31.355	33.002	34.623	36.219	37.795	39.352	40.891	42.415	43.925	45.421
	.01	35.703	37.439	39.144	40.824	42.479	44.113	45.727	47.324	48.905	50.470
54	.05	31.406	33.054	34.676	36.273	37.850	39.408	40.948	42.473	43.983	45.480
	.01	35.752	37.488	39.195	40.875	42.531	44.166	45.782	47.379	48.960	50.527
55	.05	31.456	33.105	34.728	36.326	37.904	39.463	41.004	42.530	44.041	45.539
	.01	35.800	37.537	39.245	40.926	42.583	44.219	45.835	47.433	49.015	50.582
56	.05	31.505	33.155	34.779	36.379	37.957	39.517	41.059	42.585	44.097	45.596
	.01	35.847	37.585	39.294	40.976	42.633	44.270	45.887	47.486	49.069	50.636
57	.05	31.553	33.204	34.829	36.430	38.009	39.570	41.113	42.640	44.153	45.652
	.01	35.893	37.633	39.342	41.024	42.683	44.320	45.938	47.538	49.121	50.690
58	.05	31.600	33.252	34.878	36.480	38.060	39.622	41.165	42.694	44.207	45.707
	.01	35.939	37.679	39.389	41.072	42.732	44.370	45.988	47.589	49.173	50.742
59	.05	31.646	33.300	34.926	36.529	38.110	39.673	41.217	42.746	44.261	45.762
	.01	35.983	37.724	39.435	41.120	42.780	44.419	46.038	47.639	49.224	50.794
60	.05	31.692	33.346	34.974	36.577	38.160	39.723	41.268	42.798	44.313	45.815
	.01	36.027	37.769	39.481	41.166	42.827	44.466	46.086	47.688	49.274	50.844

TABLE **F**   Critical values of the F-distribution

This table furnished critical values of the F-distribution for degrees of freedom $v_1 = 1$ to 12 in increments of one and for $v_1 = 15, 20, 24, 30, 40, 50, 60, 120$, and ∞, arranged across the top of the table, as well as for degrees of freedom $v_2 = 1$ to 30 in increments of one and for $v_2 = 40, 60, 120$, and ∞, arranged along the left margin of the table. The percentage points (corresponding to $\alpha = 0.75$, 0.50, 0.25, 0.10, 0.05, 0.025, 0.01, 0.005 and 0.001) represent the area to the right of the critical value of F in one tail of the distribution, as shown in the accompanying figure. The critical values of F are given to at least three significant figures and in a few cases to four.

To find the critical value of F for a given pair of numbers of degrees of freedom v_1 and v_2, look up these two arguments across the top of the table and in the leftmost column, respectively, and read off the desired value of F from the appropriate row in the block of nine values for the various percentage points. For example, for $v_1 = 6$ and $v_2 = 28$, $F_{.05} = 2.45$ and $F_{.001} = 5.24$. The last value indicates that 0.1% of the area under the F-distribution for $v_1 = 6$ and $v_2 = 28$ degrees of freedom is to the right of the value $F = 5.24$. This is the correct value for a one-tailed test in which the alternative hypothesis is $\sigma_1^2 > \sigma_2^2$. For a two-tailed test one divides the greater variance by the smaller one; therefore, for a type I error of α one has to look up the F-value for $\alpha/2$ in the table. Only one F-value for $\alpha > 0.5$ is given ($\alpha = 0.75$). For other values of $\alpha > 0.5$, make use of the relation $F_{\alpha[v_1,v_2]} = 1/F_{(1-\alpha)[v_2,v_1]}$. For example, one can obtain $F_{.95[8,24]}$ as $1/F_{.05[24,8]} = 1/3.12 = 0.3205$.

For a number of degrees of freedom not furnished in the arguments, employ harmonic interpolation. If both v_1 and v_2 require interpolation, you must interpolate for each of these arguments in turn. Thus, to obtain $F_{.05[55,80]}$, first interpolate between $F_{.05[50,60]}$ and $F_{.05[60,60]}$ and between $F_{.05[50,120]}$ and $F_{.05[60,120]}$ to estimate $F_{.05[55,60]}$ and $F_{.05[55,120]}$, respectively. Then interpolate between these two values to obtain the desired quantity.

There are numerous applications of the F-distribution in statistics. Its most common use is in tests of significance in analysis of variance (Chapters 8 to 12); it is also used to test the significance of differences between two variances (Section 8.3).

The values in this table were obtained in a variety of ways. Entries for $\alpha = 0.50$, 0.25, 0.10, 0.05, 0.025, 0.01, and 0.005; for $v_1 = 1$ to 10, 12, 15, 20, 24, 30, 40, 60, 120, and ∞; and for $v_2 = 1$ to 30, 40, 60, 120, and ∞ were copied from a table by M. Merrington and C. M. Thompson (*Biometrika* **33**:73–88, 1943) with permission of the publisher. F-values for $\alpha = 0.001$ and the above values of v_1 and v_2 were copied from table 18 in E. S. Pearson and H. O. Hartley,

Biometrika Tables for Statisticians, Vol. 1 (Cambridge University Press, 1958) with permission of the publishers. Entries for $v_1 = 11$ and 50 and for $v_2 = 1$ to 11 were obtained by interpolation. *F*-values for $\alpha = 0.75$, for $v_1 = 1$ to 10, 12, 15, 20, 24, 30, 40, 60, 120, and ∞, and for v_2 equal to the same numbers of degrees of freedom were obtained by computation of reciprocals of other entries in the table, as shown above. All remaining values of *F* were computed by the Cornish–Fisher method as described in section 16.21 in M. G. Kendall and A. Stuart, *The Advanced Theory of Statistics,* Vol. 1, 6th ed. (Charles Griffin, London, 1958).

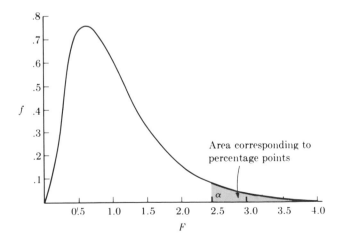

TABLE F   Critical values of the F-distribution

ν_1

(degrees of freedom of numerator mean squares)

	α	1	2	3	4	5	6
1	.75	.172	.389	.494	.553	.591	.617
	.50	1.00	1.50	1.71	1.82	1.89	1.94
	.25	5.83	7.50	8.20	8.58	8.82	8.98
	.10	39.9	49.5	53.6	55.8	57.2	58.2
	.05	161	199	216	225	230	234
	.025	648	800	864	900	922	937
	.01	4050	5000	5400	5620	5760	5860
	.005	16200	20000	21600	22500	23100	23400
	.001	405300	500000	540400	562500	576400	585900
2	.75	.133	.333	.439	.500	.540	.568
	.50	.667	1.00	1.13	1.21	1.25	1.28
	.25	2.57	3.00	3.15	3.23	3.28	3.31
	.10	8.53	9.00	9.16	9.24	9.29	9.33
	.05	18.5	19.0	19.2	19.2	19.3	19.3
	.025	38.5	39.0	39.2	39.2	39.3	39.3
	.01	98.5	99.0	99.2	99.2	99.3	99.3
	.005	198	199	199	199	199	199
	.001	999	999	999	999	999	999
3	.75	.122	.317	.424	.489	.531	.561
	.50	.585	.881	1.00	1.06	1.10	1.13
	.25	2.02	2.28	2.36	2.39	2.41	2.42
	.10	5.54	5.46	5.39	5.34	5.31	5.28
	.05	10.1	9.55	9.28	9.12	9.01	8.94
	.025	17.4	16.0	15.4	15.1	14.9	14.7
	.01	34.1	30.8	29.5	28.7	28.2	27.9
	.005	55.6	49.8	47.5	46.2	45.4	44.8
	.001	167	149	141	137	135	133
4	.75	.117	.309	.418	.484	.528	.560
	.50	.549	.828	.941	1.00	1.04	1.06
	.25	1.81	2.00	2.05	2.06	2.07	2.08
	.10	4.54	4.32	4.19	4.11	4.05	4.01
	.05	7.71	6.94	6.59	6.39	6.26	6.16
	.025	12.2	10.6	9.98	9.60	9.36	9.20
	.01	21.2	18.0	16.7	16.0	15.5	15.2
	.005	31.3	26.3	24.3	23.2	22.4	22.0
	.001	74.1	61.3	56.2	53.4	51.7	50.5
5	.75	.113	.305	.415	.483	.528	.560
	.50	.528	.799	.907	.965	1.00	1.02
	.25	1.69	1.85	1.88	1.89	1.89	1.89
	.10	4.06	3.78	3.62	3.52	3.45	3.40
	.05	6.61	5.79	5.41	5.19	5.05	4.95
	.025	10.0	8.43	7.76	7.39	7.15	6.98
	.01	16.3	13.3	12.1	11.4	11.0	10.7
	.005	22.8	18.3	16.5	15.6	14.9	14.5
	.001	47.2	37.1	33.2	31.1	29.7	28.8

ν_2 (degrees of freedom of denominator mean squares)

ν_1

(degrees of freedom of numerator mean squares)

7	8	9	10	11	12	α	
.636	.650	.661	.670	.680	.684	.75	
1.98	2.00	2.03	2.04	2.06	2.07	.50	
9.10	9.19	9.26	9.32	9.37	9.41	.25	
58.9	59.4	59.9	60.2	60.5	60.7	.10	
237	239	241	241	243	244	.05	**1**
948	957	963	969	973	977	.025	
5930	5980	6020	6060	6080	6110	.01	
23700	23900	24100	24200	24300	24400	.005	
592900	598100	602300	605600	608400	610700	.001	
.588	.604	.616	.626	.633	.641	.75	
1.30	1.32	1.33	1.35	1.36	1.36	.50	
3.34	3.35	3.37	3.38	3.39	3.39	.25	
9.35	9.37	9.38	9.39	9.40	9.41	.10	
19.4	19.4	19.4	19.4	19.4	19.4	.05	**2**
39.4	39.4	39.4	39.4	39.4	39.4	.025	
99.4	99.4	99.4	99.4	99.4	99.4	.01	
199	199	199	199	199	199	.005	
999	999	999	999	999	999	.001	
.581	.600	.613	.624	.633	.641	.75	
1.15	1.16	1.17	1.18	1.19	1.20	.50	
2.43	2.44	2.44	2.44	2.45	2.45	.25	
5.27	5.25	5.24	5.23	5.22	5.22	.10	
8.89	8.85	8.81	8.79	8.76	8.74	.05	**3**
14.6	14.5	14.5	14.4	14.3	14.3	.025	
27.7	27.5	27.3	27.2	27.1	27.1	.01	
44.4	44.1	43.9	43.7	43.5	43.4	.005	
132	131	130	129	128	128	.001	
.583	.601	.615	.627	.637	.645	.75	
1.08	1.09	1.10	1.11	1.12	1.13	.50	
2.08	2.08	2.08	2.08	2.08	2.08	.25	
3.98	3.95	3.94	3.92	3.91	3.90	.10	
6.09	6.04	6.00	5.96	5.93	5.91	.05	**4**
9.07	8.98	8.90	8.84	8.79	8.75	.025	
15.0	14.8	14.7	14.5	14.4	14.4	.01	
21.6	21.4	21.1	21.0	20.8	20.7	.005	
49.7	49.0	48.5	48.1	47.7	47.4	.001	
.584	.604	.618	.631	.641	.650	.75	
1.04	1.05	1.06	1.07	1.08	1.09	.50	
1.89	1.89	1.89	1.89	1.89	1.89	.25	
3.37	3.34	3.32	3.30	3.28	3.27	.10	
4.88	4.82	4.77	4.74	4.71	4.68	.05	**5**
6.85	6.76	6.68	6.62	6.57	6.52	.025	
10.5	10.3	10.2	10.1	9.99	9.89	.01	
14.2	14.0	13.8	13.6	13.5	13.4	.005	
28.2	27.6	27.2	26.9	26.6	26.4	.001	

ν_2 (degrees of freedom of denominator mean squares)

TABLE F   Critical values of the F-distribution (*continued*)

ν_1

(degrees of freedom of numerator mean squares)

	α	1	2	3	4	5	6
	.75	.111	.302	.413	.481	.524	.561
	.50	.515	.780	.886	.942	.977	1.00
	.25	1.62	1.76	1.78	1.79	1.79	1.78
	.10	3.78	3.46	3.29	3.18	3.11	3.05
6	.05	5.99	5.14	4.76	4.53	4.39	4.28
	.025	8.81	7.26	6.60	6.23	5.99	5.82
	.01	13.7	10.9	9.78	9.15	8.75	8.47
	.005	18.6	14.5	12.9	12.0	11.5	11.1
	.001	35.5	27.0	23.7	21.9	20.8	20.0
	.75	.110	.300	.412	.481	.528	.562
	.50	.506	.767	.871	.926	.960	.983
	.25	1.57	1.70	1.72	1.72	1.71	1.71
	.10	3.59	3.26	3.07	2.96	2.88	2.83
7	.05	5.59	4.74	4.35	4.12	3.97	3.87
	.025	8.07	6.54	5.89	5.52	5.29	5.12
	.01	12.2	9.55	8.45	7.85	7.46	7.19
	.005	16.2	12.4	10.9	10.1	9.52	9.16
	.001	29.3	21.7	18.8	17.2	16.2	15.5
	.75	.109	.298	.411	.481	.529	.563
	.50	.499	.757	.860	.915	.948	.971
	.25	1.54	1.66	1.67	1.66	1.66	1.65
	.10	3.46	3.11	2.92	2.81	2.73	2.67
8	.05	5.32	4.46	4.07	3.84	3.69	3.58
	.025	7.57	6.06	5.42	5.05	4.82	4.65
	.01	11.3	8.65	7.59	7.01	6.63	6.37
	.005	14.7	11.0	9.60	8.81	8.30	7.95
	.001	25.4	18.5	15.8	14.4	13.5	12.9
	.75	.108	.297	.410	.480	.529	.564
	.50	.494	.749	.852	.906	.939	.962
	.25	1.51	1.62	1.63	1.63	1.62	1.61
	.10	3.36	3.01	2.81	2.69	2.61	2.55
9	.05	5.12	4.26	3.86	3.63	3.48	3.37
	.025	7.21	5.71	5.08	4.72	4.48	4.32
	.01	10.6	8.02	6.99	6.42	6.06	5.80
	.005	13.6	10.1	8.72	7.96	7.47	7.13
	.001	22.9	16.4	13.9	12.6	11.7	11.1
	.75	.107	.296	.409	.480	.529	.565
	.50	.490	.743	.845	.899	.932	.954
	.25	1.49	1.60	1.60	1.59	1.59	1.58
	.10	3.29	2.92	2.73	2.61	2.52	2.46
10	.05	4.96	4.10	3.71	3.48	3.33	3.22
	.025	6.94	5.46	4.83	4.47	4.24	4.07
	.01	10.0	7.56	6.55	5.99	5.64	5.39
	.005	12.8	9.43	8.08	7.34	6.87	6.54
	.001	21.0	14.9	12.5	11.3	10.5	9.92

ν_2 (degrees of freedom of denominator mean squares)

ν_1

(degrees of freedom of numerator mean squares)

7	8	9	10	11	12	α	
.586	.606	.621	.635	.645	.654	.75	
1.02	1.03	1.04	1.05	1.06	1.06	.50	
1.78	1.78	1.77	1.77	1.77	1.77	.25	
3.01	2.98	2.96	2.94	2.92	2.90	.10	
4.21	4.15	4.10	4.06	4.03	4.00	.05	**6**
5.70	5.60	5.52	5.46	5.41	5.37	.025	
8.26	8.10	7.98	7.87	7.79	7.72	.01	
10.8	10.6	10.4	10.3	10.1	10.0	.005	
19.5	19.0	18.7	18.4	18.2	18.0	.001	
.588	.608	.624	.637	.649	.658	.75	
1.00	1.01	1.02	1.03	1.04	1.04	.50	
1.70	1.70	1.69	1.69	1.68	1.68	.25	
2.78	2.75	2.72	2.70	2.68	2.67	.10	
3.77	3.73	3.68	3.64	3.60	3.57	.05	**7**
4.99	4.89	4.82	4.76	4.71	4.67	.025	
6.99	6.84	6.72	6.62	6.54	6.47	.01	
8.89	8.68	8.52	8.38	8.27	8.18	.005	
15.0	14.6	14.3	14.1	13.9	13.7	.001	
.589	.610	.627	.640	.654	.661	.75	
.988	1.00	1.01	1.02	1.03	1.03	.50	
1.64	1.64	1.63	1.63	1.63	1.62	.25	
2.62	2.59	2.56	2.54	2.52	2.50	.10	
3.50	3.44	3.39	3.35	3.31	3.28	.05	**8**
4.53	4.43	4.36	4.30	4.25	4.20	.025	
6.18	6.03	5.91	5.81	5.73	5.67	.01	
7.69	7.50	7.34	7.21	7.10	7.01	.005	
12.4	12.0	11.8	11.5	11.3	11.2	.001	
.591	.612	.629	.643	.654	.664	.75	
.978	.990	1.00	1.01	1.02	1.02	.50	
1.60	1.60	1.59	1.59	1.58	1.58	.25	
2.51	2.47	2.44	2.42	2.40	2.38	.10	
3.29	3.23	3.18	3.14	3.10	3.07	.05	**9**
4.20	4.10	4.03	3.96	3.91	3.87	.025	
5.61	5.47	5.35	5.26	5.18	5.11	.01	
6.88	6.69	6.54	6.42	6.32	6.23	.005	
10.7	10.4	10.1	9.79	9.72	9.57	.001	
.592	.613	.631	.645	.657	.667	.75	
.971	.983	.992	1.00	1.01	1.01	.50	
1.57	1.56	1.56	1.55	1.54	1.54	.25	
2.41	2.38	2.35	2.32	2.30	2.28	.10	
3.14	3.07	3.02	2.98	2.94	2.91	.05	**10**
3.95	3.85	3.78	3.72	3.67	3.62	.025	
5.20	5.06	4.94	4.85	4.77	4.71	.01	
6.30	6.12	5.97	5.85	5.75	5.66	.005	
9.52	9.20	8.96	8.75	8.59	8.45	.001	

ν_2 (degrees of freedom of denominator mean squares)

TABLE F   Critical values of the *F*-distribution (*continued*)

v_1

(degrees of freedom of numerator mean squares)

	α	1	2	3	4	5	6
	.75	.107	.295	.408	.481	.529	.565
	.50	.486	.739	.840	.893	.926	.948
	.25	1.47	1.58	1.58	1.57	1.56	1.55
	.10	3.23	2.86	2.66	2.54	2.45	2.39
11	.05	4.84	3.98	3.59	3.36	3.20	3.09
	.025	6.72	5.26	4.63	4.28	4.04	3.88
	.01	9.65	7.21	6.22	5.67	5.32	5.07
	.005	12.2	8.91	7.60	6.88	6.42	6.10
	.001	19.7	13.8	11.6	10.3	9.58	9.05
	.75	.106	.295	.408	.480	.530	.566
	.50	.484	.735	.835	.888	.921	.943
	.25	1.46	1.56	1.56	1.55	1.54	1.53
	.10	3.18	2.81	2.61	2.48	2.39	2.33
12	.05	4.75	3.89	3.49	3.26	3.11	3.00
	.025	6.55	5.10	4.47	4.12	3.89	3.73
	.01	9.33	6.93	5.95	5.41	5.06	4.82
	.005	11.8	8.51	7.23	6.52	6.07	5.76
	.001	18.6	13.0	10.8	9.63	8.89	8.38
	.75	.106	.294	.408	.480	.530	.567
	.50	.481	.731	.832	.885	.917	.939
	.25	1.45	1.55	1.55	1.53	1.52	1.51
	.10	3.14	2.76	2.56	2.43	2.35	2.28
13	.05	4.67	3.81	3.41	3.18	3.03	2.92
	.025	6.41	4.97	4.35	4.00	3.77	3.60
	.01	9.07	6.70	5.74	4.21	4.86	4.62
	.005	11.4	8.19	6.93	6.23	5.79	5.48
	.001	17.8	12.3	10.2	9.07	8.35	7.86
	.75	.106	.294	.408	.480	.530	.567
	.50	.479	.729	.828	.881	.914	.936
	.25	1.44	1.53	1.53	1.52	1.51	1.50
	.10	3.10	2.73	2.52	2.39	2.31	2.24
14	.05	4.60	3.74	3.34	3.11	2.96	2.85
	.025	6.30	4.86	4.24	3.89	3.66	3.50
	.01	8.86	6.51	5.56	5.04	4.69	4.46
	.005	11.1	7.92	6.68	6.00	5.53	5.26
	.001	17.1	11.8	9.73	8.62	7.92	7.43
	.75	.105	.293	.407	.480	.530	.568
	.50	.478	.726	.826	.878	.911	.933
	.25	1.43	1.52	1.52	1.51	1.49	1.48
	.10	3.07	2.70	2.49	2.36	2.27	2.21
15	.05	4.54	3.68	3.29	3.06	2.90	2.79
	.025	6.20	4.77	4.15	3.80	3.58	3.41
	.01	8.68	6.36	5.42	4.89	4.56	4.32
	.005	10.8	7.70	6.48	5.80	5.37	5.07
	.001	16.6	11.3	9.34	8.25	7.57	7.09

v_2 (degrees of freedom of denominator mean squares)

ν_1
(degrees of freedom of numerator mean squares)

7	8	9	10	11	12	α	
.592	.614	.633	.645	.658	.667	.75	
.964	.977	.986	.994	1.00	1.01	.50	
1.54	1.53	1.53	1.52	1.51	1.51	.25	
2.34	2.30	2.27	2.25	2.23	2.21	.10	
3.01	2.95	2.90	2.85	2.82	2.79	.05	**11**
3.76	3.66	3.59	3.53	3.48	3.43	.025	
4.89	4.74	4.63	4.54	4.46	4.40	.01	
5.86	5.68	5.54	5.42	5.32	5.24	.005	
8.66	8.35	8.12	7.92	7.76	7.63	.001	
.594	.616	.633	.649	.661	.671	.75	
.959	.972	.981	.989	.995	1.00	.50	
1.52	1.51	1.51	1.50	1.50	1.49	.25	
2.28	2.24	2.21	2.19	2.17	2.15	.10	
2.91	2.85	2.80	2.75	2.72	2.69	.05	**12**
3.61	3.51	3.44	3.37	3.32	3.28	.025	
4.64	4.50	4.39	4.30	4.22	4.16	.01	
5.52	5.35	5.20	5.09	4.99	4.91	.005	
8.00	7.71	7.48	7.29	7.14	7.00	.001	
.595	.617	.635	.650	.662	.673	.75	
.955	.967	.977	.984	.990	.996	50	
1.50	1.49	1.49	1.48	1.48	1.47	.25	
2.23	2.20	2.16	2.14	2.12	2.10	.10	
2.83	2.77	2.71	2.67	2.64	2.60	.05	**13**
3.48	3.39	3.31	3.25	3.20	3.15	.025	
4.44	4.30	4.19	4.10	4.03	3.96	.01	
5.25	5.08	4.94	4.82	4.73	4.64	.005	
7.49	7.21	6.98	6.80	6.65	6.52	.001	
.595	.618	.636	.651	.664	.674	.75	
.952	.964	.973	.981	.987	.992	.50	
1.49	1.48	1.47	1.46	1.46	1.45	.25	
2.19	2.45	1.42	2.10	2.07	2.05	.10	
2.76	2.70	2.65	2.60	2.57	2.53	.05	**14**
3.38	3.29	3.20	3.15	3.10	3.05	.025	
4.28	4.14	4.03	3.94	3.86	3.80	.01	
5.03	4.86	4.72	4.60	4.51	4.43	.005	
7.08	6.80	6.58	6.40	6.26	6.13	.001	
.596	.618	.637	.652	.667	.676	.75	
.948	.960	.970	.977	.983	.989	.50	
1.47	1.46	1.46	1.45	1.44	1.44	.25	
2.16	2.12	2.09	2.06	2.04	2.02	.10	
2.71	2.64	2.59	2.54	2.51	2.48	.05	**15**
3.29	3.20	3.12	3.06	3.01	2.96	.025	
4.14	4.00	3.89	3.80	3.73	3.67	.01	
4.85	4.67	4.54	4.42	4.33	4.25	.005	
6.74	6.47	6.26	6.08	5.94	5.81	.001	

ν_2 (degrees of freedom of denominator mean squares)

TABLE F Critical values of the F-distribution (*continued*)

ν_1

(degrees of freedom of numerator mean squares)

	α	1	2	3	4	5	6
	.75	.105	.293	.407	.480	.531	.568
	.50	.476	.724	.823	.876	.908	.930
	.25	1.42	1.51	1.51	1.50	1.48	1.47
	.10	3.05	2.67	2.46	2.33	2.24	2.18
16	.05	4.49	3.63	3.24	3.01	2.85	2.74
	.025	6.12	4.69	4.08	3.73	3.50	3.34
	.01	8.53	6.23	5.29	4.77	4.44	4.20
	.005	10.6	7.51	6.30	5.64	5.21	4.91
	.001	16.1	11.0	9.00	7.94	7.27	6.81
	.75	.105	.292	.407	.480	.531	.568
	.50	.475	.722	.821	.874	.906	.928
	.25	1.42	1.51	1.50	1.49	1.47	1.46
	.10	3.03	2.64	2.44	2.31	2.22	2.15
17	.05	4.45	3.59	3.20	2.96	2.91	2.70
	.025	6.04	4.62	4.01	3.66	3.44	3.28
	.01	8.40	6.11	5.19	4.67	4.34	4.10
	.005	10.4	7.35	6.16	5.50	5.07	4.78
	.001	15.7	10.7	8.73	7.68	7.02	6.56
	.75	.105	.292	.407	.480	.531	.569
	.50	.474	.721	.819	.872	.904	.926
	.25	1.41	1.50	1.49	1.48	1.46	1.45
	.10	3.01	2.62	2.42	2.29	2.20	2.13
18	.05	4.41	3.55	3.16	2.93	2.77	2.66
	.025	5.98	4.56	3.95	3.61	3.38	3.22
	.01	8.28	6.01	5.09	4.58	4.25	4.01
	.005	10.2	7.21	6.03	5.37	4.96	4.66
	.001	15.4	10.4	8.49	7.46	6.81	6.35
	.75	.104	.292	.407	.480	.531	.569
	.50	.473	.719	.818	.870	.902	.924
	.25	1.41	1.49	1.49	1.47	1.46	1.44
	.10	2.99	2.61	2.40	2.27	2.18	2.11
19	.05	4.38	3.52	3.13	2.90	2.74	2.63
	.025	5.92	4.51	3.90	3.56	3.33	3.17
	.01	8.19	5.93	5.01	4.50	4.17	3.94
	.005	10.1	7.09	5.92	5.27	4.85	4.56
	.001	15.1	10.2	8.28	7.26	6.62	6.18
	.75	.104	.292	.407	.480	.531	.569
	.50	.472	.718	.816	.868	.900	.922
	.25	1.40	1.49	1.48	1.47	1.45	1.44
	.10	2.97	2.59	2.38	2.25	2.16	2.09
20	.05	4.35	3.49	3.10	2.87	2.71	2.60
	.025	5.87	4.46	3.86	3.51	3.29	3.13
	.01	8.10	5.85	4.94	4.43	4.10	3.87
	.005	9.94	6.99	5.82	5.17	4.76	4.47
	.001	14.8	9.95	8.10	7.10	6.46	6.02

ν_2 (degrees of freedom of denominator mean squares)

ν_1

(degrees of freedom of numerator mean squares)

7	8	9	10	11	12	α	
.597	.619	.638	.653	.666	.677	.75	
.946	.958	.967	.975	.981	.986	.50	
1.46	1.45	1.44	1.44	1.43	1.43	.25	
2.13	2.09	2.06	2.03	2.01	1.99	.10	
2.66	2.59	2.54	2.49	2.46	2.42	.05	**16**
3.22	3.12	3.05	2.99	2.93	2.89	.025	
4.03	3.89	3.78	3.69	3.62	3.55	.01	
4.69	4.52	5.38	4.27	4.18	4.10	.005	
6.46	6.19	5.98	5.81	5.67	5.55	.001	
.597	.620	.639	.654	.667	.678	.75	
.943	.955	.965	.972	.978	.983	.50	
1.45	1.44	1.43	1.43	1.42	1.41	.25	
2.10	2.06	2.03	2.00	1.98	1.96	.10	
3.61	2.55	2.49	2.45	2.41	2.38	.05	**17**
3.16	3.06	2.98	2.92	2.87	2.82	.025	
3.93	3.79	3.68	3.59	3.52	3.46	.01	
4.56	4.39	4.25	4.14	4.05	3.97	.005	
6.22	5.96	5.75	5.58	5.44	5.32	.001	
.598	.621	.639	.655	.668	.679	.75	
.941	.953	.962	.970	.976	.981	.50	
1.44	1.43	1.42	1.42	1.41	1.40	.25	
2.08	2.04	2.00	1.98	1.95	1.93	.10	
2.58	2.51	2.46	2.41	2.37	2.34	.05	**18**
3.10	3.01	2.93	2.87	2.81	2.77	.025	
3.84	3.71	3.60	3.51	3.43	3.37	.01	
4.44	4.28	4.14	4.03	3.94	3.86	.005	
6.02	5.76	5.56	5.39	5.25	5.13	.001	
.598	.621	.640	.656	.669	.680	.75	
.939	.951	.961	.968	.974	.979	.50	
1.43	1.42	1.41	1.41	1.40	1.40	.25	
2.06	2.02	1.98	1.96	1.93	1.91	.10	
2.54	2.48	2.42	2.38	2.34	2.31	.05	**19**
3.05	2.76	2.88	2.82	2.76	2.72	.025	
3.77	3.63	3.52	3.43	3.56	3.30	.01	
4.34	4.18	4.04	3.93	3.84	3.76	.005	
5.85	5.59	5.39	5.22	5.08	4.97	.001	
.598	.622	.641	.656	.671	.681	.75	
.938	.950	.959	.966	.972	.977	.50	
1.43	1.42	1.41	1.40	1.39	1.39	.25	
2.04	2.00	1.96	1.94	1.91	1.89	.10	
2.51	2.45	2.39	2.35	2.31	2.28	.05	**20**
3.01	2.91	2.84	2.77	2.72	2.68	.025	
3.70	3.56	3.46	3.37	3.29	3.23	.01	
4.26	4.09	3.96	3.85	3.76	3.68	.005	
5.69	5.44	5.24	5.08	4.94	4.82	.001	

ν_2 (degrees of freedom of denominator mean squares)

TABLE F   Critical values of the *F*-distribution (*continued*)

ν_1

(degrees of freedom of numerator mean squares)

ν_2 (degrees of freedom of denominator mean squares)

	α	1	2	3	4	5	6
21	.75	.104	.292	.407	.480	.532	.570
	.50	.471	.717	.815	.867	.899	.921
	.25	1.40	1.48	1.48	1.46	1.44	1.43
	.10	2.96	2.57	2.36	2.23	2.14	2.08
	.05	4.32	3.47	3.07	2.84	2.68	2.57
	.025	5.83	4.42	3.82	3.48	3.25	3.09
	.01	8.02	5.75	4.87	4.37	4.04	3.81
	.005	9.83	6.89	5.73	5.09	4.99	4.39
	.001	14.6	9.77	7.94	6.95	6.32	5.88
22	.75	.104	.292	.407	.481	.532	.570
	.50	.470	.715	.814	.866	.898	.919
	.25	1.40	1.48	1.47	1.45	1.44	1.42
	.10	2.95	2.56	2.35	2.22	2.13	2.06
	.05	4.30	3.44	3.05	2.82	2.66	2.55
	.025	5.79	4.38	3.78	3.44	3.22	3.05
	.01	7.95	5.72	4.82	4.31	3.99	3.76
	.005	6.73	6.81	5.65	5.02	4.61	4.32
	.001	14.4	9.61	7.80	6.81	6.19	5.76
23	.75	.104	.291	.406	.481	.532	.570
	.50	.470	.714	.813	.864	.896	.918
	.25	1.39	1.47	1.47	1.45	1.43	1.42
	.10	2.94	2.55	2.31	2.25	2.11	1.05
	.05	4.28	3.42	3.03	2.80	2.64	2.53
	.025	5.75	4.35	3.75	3.41	3.18	3.02
	.01	7.88	5.66	4.76	4.26	3.94	3.71
	.005	9.63	6.73	5.58	4.95	4.54	4.26
	.001	14.2	9.47	7.67	6.69	6.08	5.65
24	.75	.104	.291	.406	.480	.532	.570
	.50	.469	.714	.812	.863	.895	.917
	.25	1.39	1.47	1.46	1.44	1.43	1.41
	.10	2.93	2.54	2.33	2.19	2.10	2.04
	.05	4.26	3.40	3.01	2.78	2.62	2.51
	.025	5.72	4.32	3.72	3.38	3.15	2.99
	.01	7.82	5.61	4.72	4.22	3.90	3.67
	.005	9.55	6.66	5.52	4.89	4.49	4.20
	.001	14.0	9.34	7.55	6.59	5.98	5.55
25	.75	.104	.292	.407	.481	.532	.571
	.50	.468	.713	.811	.862	.984	.916
	.25	1.39	1.47	1.46	1.44	1.42	1.41
	.10	2.92	2.53	2.32	2.18	2.09	2.02
	.05	4.24	3.39	2.99	2.76	2.60	2.49
	.025	5.69	4.29	3.69	3.35	3.13	2.97
	.01	7.77	5.57	4.68	4.18	3.86	3.63
	.005	9.48	6.60	5.46	4.84	4.43	4.15
	.001	13.9	9.22	7.45	6.49	5.88	5.46

ν_1

(degrees of freedom of numerator mean squares)

7	8	9	10	11	12	α	ν_2
.599	.622	.641	.657	.671	.682	.75	
.936	.948	.957	.965	.971	.976	.50	
1.42	1.41	1.40	1.39	1.39	1.38	.25	
2.02	1.98	1.95	1.92	1.90	1.87	.10	
2.49	2.42	2.37	2.32	2.28	2.25	.05	**21**
2.97	2.87	2.80	2.73	2.68	2.64	.025	
3.64	3.51	3.40	3.31	3.24	3.17	.01	
4.18	4.01	3.88	3.77	3.68	3.60	.005	
5.56	5.31	5.11	4.95	4.81	4.70	.001	
.599	.623	.642	.658	.671	.683	.75	
.935	.947	.956	.963	.969	.974	.50	
1.41	1.40	1.39	1.39	1.38	1.37	.25	
2.01	1.97	1.93	1.90	1.88	1.86	.10	
2.46	2.40	2.39	2.30	2.26	2.23	.05	**22**
2.93	2.84	2.76	2.70	2.65	2.60	.025	
3.59	3.45	3.35	3.26	3.18	3.12	.01	
4.11	3.94	6.81	3.70	3.61	3.53	.005	
5.44	5.19	4.99	4.83	4.70	4.58	.001	
.600	.623	.642	.658	.672	.684	.75	
.934	.945	.955	.962	.968	.973	.50	
1.41	1.40	1.39	1.38	1.37	1.37	.25	
1.99	1.95	1.92	1.89	1.87	1.85	.10	
2.44	2.37	2.32	2.27	2.24	2.20	.05	**23**
2.90	2.81	2.73	2.67	2.62	2.57	.025	
3.54	3.41	3.30	3.21	3.14	3.07	.01	
4.05	3.88	3.75	3.64	3.55	3.47	.005	
5.33	5.09	4.89	4.73	4.59	4.48	.001	
.600	.623	.643	.659	.671	.684	.75	
.932	.944	.953	.961	.967	.972	.50	
1.40	1.39	1.38	1.37	1.37	1.36	.25	
1.98	1.94	1.91	1.88	1.85	1.83	.10	
2.42	2.36	2.30	2.25	2.22	2.18	.05	**24**
2.87	2.78	2.70	2.64	2.59	2.54	.025	
3.50	3.36	3.26	3.17	3.09	3.03	.01	
3.99	3.83	3.69	3.59	3.50	3.42	.005	
5.23	4.99	4.80	4.64	4.50	4.39	.001	
.600	.624	.643	.659	.673	.685	.75	
.931	.943	.952	.960	.966	.971	.50	
1.40	1.39	1.38	1.37	1.36	1.36	.25	
1.97	1.93	1.89	1.87	1.84	1.82	.10	
2.40	2.34	2.28	2.24	2.20	2.16	.05	**25**
2.85	2.75	2.68	2.61	2.56	2.51	.025	
3.46	3.32	3.22	3.13	3.06	2.99	.01	
3.94	3.78	3.64	3.54	3.44	3.37	.005	
5.15	4.91	4.71	4.56	4.42	4.31	.001	

ν_2 (degrees of freedom of denominator mean squares)

TABLE F Critical values of the *F*-distribution (*continued*)

ν_1

(degrees of freedom of numerator mean squares)

	α	1	2	3	4	5	6
	.75	.104	.291	.406	.480	.532	.571
	.50	.468	.712	.810	.861	.893	.915
	.25	1.38	1.46	1.45	1.44	1.42	1.41
	.10	2.91	2.52	2.31	2.17	2.08	2.01
26 .05		4.23	3.37	2.98	2.74	2.59	2.47
	.025	5.66	4.27	3.67	3.33	3.10	2.94
	.01	7.72	5.53	4.64	4.14	3.82	3.59
	.005	9.41	6.54	5.41	4.79	4.38	4.10
	.001	13.7	9.12	7.36	6.41	5.80	5.38
	.75	.104	.291	.406	.480	.532	.571
	.50	.467	.711	.809	.861	.892	.914
	.25	1.38	1.46	1.45	1.43	1.42	1.40
	.10	2.90	2.51	2.30	2.17	2.07	2.00
27 .05		4.21	3.35	2.96	2.73	2.57	2.46
	.025	5.63	4.24	3.65	3.31	3.08	2.92
	.01	7.68	5.49	4.60	4.11	3.78	3.56
	.005	9.34	6.49	5.36	4.74	4.34	4.06
	.001	13.6	9.02	7.27	6.33	5.73	5.31
	.75	.103	.290	.406	.480	.532	.571
	.50	.467	.711	.808	.860	.892	.913
	.25	1.38	1.46	1.45	1.43	1.41	1.40
	.10	2.89	2.50	2.29	2.16	2.06	2.00
28 .05		4.20	3.34	2.95	2.71	2.56	2.45
	.025	5.61	4.22	3.63	3.29	3.06	2.90
	.01	7.64	5.45	4.57	4.07	3.75	3.53
	.005	9.28	6.44	5.32	4.70	4.30	4.02
	.001	13.5	8.93	7.19	6.25	5.66	5.24
	.75	.103	.290	.406	.480	.532	.571
	.50	.467	.710	.808	.859	.891	.912
	.25	1.38	1.45	1.45	1.43	1.41	1.40
	.10	2.89	2.50	2.28	2.15	2.06	1.99
29 .05		4.18	3.33	2.93	2.70	2.55	2.43
	.025	5.59	4.20	3.60	3.27	3.04	2.88
	.01	7.60	5.42	4.54	4.04	3.73	3.50
	.005	9.23	6.40	5.28	4.66	4.26	3.98
	.001	13.4	8.85	7.12	6.19	5.59	5.18
	.75	.103	.290	.406	.480	.532	.571
	.50	.466	.709	.807	.858	.890	.912
	.25	1.38	1.45	1.44	1.42	1.41	1.39
	.10	2.88	2.49	2.28	2.14	2.05	1.98
30 .05		4.17	3.32	2.92	2.69	2.53	2.42
	.025	5.57	4.18	3.59	3.25	3.03	2.87
	.01	7.56	5.39	4.51	4.02	3.70	3.47
	.005	9.18	6.35	5.24	4.62	4.23	3.95
	.001	13.3	8.77	7.05	6.12	5.53	5.12

ν_2 (degrees of freedom of denominator mean squares)

$$\nu_1$$

(degrees of freedom of numerator mean squares)

7	8	9	10	11	12	α	
.600	.624	.644	.660	.674	.686	.75	
.930	.942	.951	.959	.965	.970	.50	
1.39	1.38	1.37	1.37	1.36	1.35	.25	
1.96	1.92	1.88	1.86	1.83	1.81	.10	
2.39	2.32	2.27	2.22	2.18	2.15	.05	**26**
2.82	2.73	2.65	2.59	2.54	2.49	.025	
3.42	3.29	3.18	3.09	3.02	2.96	.01	
3.89	3.73	3.60	3.49	3.40	3.33	.005	
5.07	4.83	4.64	4.48	4.35	4.24	.001	
.601	.624	.644	.660	.674	.686	.75	
.930	.941	.950	.958	.964	.969	.50	
1.39	1.38	1.37	1.36	1.35	1.35	.25	
1.95	1.91	1.87	1.85	1.82	1.80	.10	
2.37	2.31	2.25	2.20	2.17	2.13	.05	**27**
2.80	2.71	2.63	2.57	2.51	2.47	.025	
3.39	3.26	3.45	3.06	2.99	2.93	.01	
3.85	3.69	3.56	3.45	3.36	3.28	.005	
5.00	4.76	4.57	4.41	4.28	4.17	.001	
.601	.625	.644	.661	.675	.687	.75	
.929	.940	.950	.957	.963	.968	.50	
1.39	1.38	1.37	1.36	1.35	1.34	.25	
1.94	1.90	1.87	1.84	1.81	1.79	.10	
2.36	2.29	2.24	2.19	2.15	2.12	.05	**28**
2.78	2.69	2.61	2.55	2.49	2.45	.025	
3.36	3.23	3.12	3.03	2.96	2.90	.01	
3.81	3.65	3.52	3.41	3.32	3.25	.005	
4.93	4.69	4.50	4.35	4.22	4.11	.001	
.601	.625	.645	.661	.675	.687	.75	
.928	.940	.949	.956	.962	.967	.50	
1.38	1.37	1.36	1.35	1.35	1.34	.25	
1.93	1.89	2.86	1.83	1.80	1.78	.10	
2.35	2.28	2.22	2.18	2.14	2.10	.05	**29**
2.76	2.67	2.59	2.51	2.47	2.43	.025	
3.33	3.20	3.09	3.00	2.93	2.87	.01	
3.77	3.61	3.48	3.38	3.29	3.21	.005	
4.87	4.64	4.45	4.29	4.16	4.05	.001	
.601	.625	.645	.661	.676	.688	.75	
.927	.939	.948	.955	.961	.966	.50	
1.38	1.37	1.36	1.35	1.34	1.34	.25	
1.93	1.88	1.85	1.82	1.79	1.77	.10	
2.33	2.27	2.21	2.16	2.13	2.09	.05	**30**
2.75	2.65	2.57	2.51	2.46	2.41	.025	
3.30	3.17	3.07	2.98	2.90	2.84	.01	
3.74	3.58	3.45	3.34	3.25	3.18	.005	
4.82	4.58	4.39	4.24	4.11	4.00	.001	

ν_2 (degrees of freedom of denominator mean squares)

TABLE F Critical values of the *F*-distribution (*continued*)

ν_1

(degrees of freedom of numerator mean squares)

ν_2	α	1	2	3	4	5	6
40	.75	.103	.289	.404	.480	.533	.572
	.50	.463	.705	.802	.854	.885	.907
	.25	1.36	1.44	1.42	1.40	1.39	1.37
	.10	2.84	2.44	2.23	2.09	2.00	1.93
	.05	4.08	3.23	2.84	2.61	2.45	2.34
	.025	5.42	4.05	3.46	3.13	2.90	2.74
	.01	7.31	5.18	4.31	3.83	3.51	3.29
	.005	8.83	6.07	4.98	4.37	3.99	3.71
	.001	12.6	8.25	6.60	5.70	5.13	4.73
60	.75	.102	.289	.405	.480	.534	.573
	.50	.461	.701	.798	.849	.880	.901
	.25	1.35	1.42	1.41	1.38	1.37	1.35
	.10	2.79	2.39	2.18	2.04	1.95	1.87
	.05	4.00	3.15	2.76	2.53	2.37	2.25
	.025	5.29	3.93	3.34	3.01	2.79	2.63
	.01	7.08	4.98	4.13	3.65	3.34	3.12
	.005	8.49	5.79	4.73	4.14	3.76	3.49
	.001	12.0	7.76	6.17	5.31	4.76	4.37
120	.75	.102	.288	.405	.481	.534	.574
	.50	.458	.697	.793	.844	.875	.896
	.25	1.34	1.40	1.39	1.37	1.35	1.33
	.10	2.75	2.35	2.13	1.99	1.90	1.82
	.05	3.92	3.07	2.68	2.45	2.29	2.17
	.025	5.15	3.80	3.23	2.89	2.67	2.52
	.01	6.85	4.79	3.95	3.48	3.17	2.96
	.005	8.18	5.54	4.50	3.92	3.55	3.28
	.001	11.4	7.32	5.79	4.95	4.42	4.04
∞	.75	.102	.288	.404	.481	.535	.576
	.50	.455	.693	.789	.839	.870	.891
	.25	1.32	1.39	1.37	1.35	1.33	1.31
	.10	2.71	2.30	2.08	1.94	1.85	1.77
	.05	3.84	3.00	2.60	2.37	2.21	2.10
	.025	5.02	3.69	3.11	2.79	2.57	2.41
	.01	6.63	4.61	3.78	3.32	3.02	2.80
	.005	7.88	5.30	4.28	3.72	3.35	3.09
	.001	10.8	6.91	5.42	4.62	4.10	3.74

ν_2 (degrees of freedom of denominator mean squares)

$$\nu_1$$

(degrees of freedom of numerator mean squares)

7	8	9	10	11	12	α	
.603	.627	.647	.662	.679	.689	.75	
.922	.934	.943	.950	.956	.961	.50	
1.36	1.35	1.34	1.33	1.32	1.31	.25	
1.87	1.83	1.79	1.76	1.74	1.74	.10	
2.25	2.18	2.12	2.08	2.04	2.04	.05	**40**
2.62	2.53	2.45	2.39	2.33	2.29	.025	
3.12	2.99	2.89	2.80	2.73	2.66	.01	
3.51	3.35	3.22	3.12	3.03	2.95	.005	
4.44	4.21	4.02	3.87	3.74	3.64	.001	
.604	.629	.650	.667	.682	.695	.75	
.917	.928	.937	.945	.951	.956	.50	
1.33	1.32	1.31	1.30	1.29	1.29	.25	
1.82	1.77	1.74	1.71	1.68	1.66	.10	
2.17	2.10	2.04	1.99	1.95	1.92	.05	**60**
2.51	2.41	2.33	2.27	2.22	2.17	.025	
2.95	2.82	2.72	2.63	2.56	2.50	.01	
3.29	3.13	3.01	2.90	2.81	2.74	.005	
4.09	3.87	3.69	3.54	3.41	3.31	.001	
.606	.631	.652	.670	.686	.699	.75	
.912	.923	.932	.939	.945	.950	.50	
1.31	1.30	1.29	1.28	1.27	1.26	.25	
1.77	1.72	1.68	1.65	1.63	1.60	.10	
2.09	2.02	1.96	1.91	1.87	1.83	.05	**120**
2.39	2.30	2.22	2.16	2.10	2.05	.025	
2.79	2.66	2.56	2.47	2.40	2.34	.01	
3.09	2.93	2.81	2.71	2.62	2.54	.005	
3.77	3.55	3.38	3.24	3.11	3.02	.001	
.608	.634	.655	.674	.690	.703	.75	
.907	.918	.927	.934	.939	.945	.50	
1.29	1.28	1.27	1.25	1.24	1.24	.25	
1.72	1.67	1.63	1.60	1.57	1.55	.10	
2.01	1.94	1.88	1.83	1.79	1.75	.05	∞
2.29	2.19	2.11	2.05	1.99	1.94	.025	
2.64	2.51	2.41	2.32´	2.25	2.18	.01	
2.90	2.74	2.62	2.52	2.43	2.36	.005	
3.47	3.27	3.10	2.96	2.84	2.74	.001	

ν_2 (degrees of freedom of denominator mean squares)

TABLE F Critical values of the F-distribution (*continued*)

$$\nu_1$$
(degrees of freedom of numerator mean squares)

ν₂ (degrees of freedom of denominator mean squares)

	α	15	20	24	30	40	50	60	120	∞
1	.75	.698	.712	.719	.727	.734	.738	.741	.749	.756
	.50	2.09	2.12	2.13	2.15	2.16	2.17	2.17	2.18	2.20
	.25	9.49	9.58	9.63	9.67	9.71	9.74	9.76	9.80	9.85
	.10	61.2	61.7	62.0	62.3	62.5	62.7	62.8	63.1	63.3
	.05	246	248	249	250	251	252	252	253	254
	.025	985	993	997	1000	1010	1010	1010	1010	1020
	.01	6160	6210	6230	6260	6290	6300	6310	6340	6370
	.005	24630	24836	24940	25440	25148	25211	25253	25359	25465
	.001	615800	620900	623500	626100	628700	630300	631300	634000	636600
2	.75	.657	.672	.680	.689	.697	.702	.705	.713	.721
	.50	1.38	1.39	1.40	1.41	1.42	1.43	1.43	1.43	1.44
	.25	3.41	3.43	3.43	3.44	3.45	3.46	3.46	3.47	3.48
	.10	9.42	9.44	9.45	9.46	9.47	9.47	9.47	9.48	9.49
	.05	19.4	19.4	19.5	19.5	19.5	19.5	19.5	19.5	19.5
	.025	39.4	39.4	39.5	39.5	39.5	39.5	39.5	39.5	39.5
	.01	99.4	99.4	99.5	99.5	99.5	99.5	99.5	99.5	99.5
	.005	199	199	199	199	199	199	199	199	200
	.001	999	999	1000	1000	1000	1000	1000	1000	1000
3	.75	.658	.675	.684	.694	.702	.708	.711	.721	.730
	.50	1.21	1.23	1.23	1.24	1.25	1.25	1.25	1.26	1.27
	.25	2.46	2.46	2.46	2.47	2.47	2.47	2.47	2.47	2.47
	.10	5.20	5.18	5.18	5.17	5.16	5.15	5.15	5.14	5.13
	.05	8.70	8.66	8.64	8.62	8.59	8.58	8.57	8.55	8.53
	.025	14.3	14.2	14.1	14.1	14.0	14.0	14.0	13.9	13.9
	.01	26.9	26.7	26.6	26.5	26.4	26.3	26.3	26.2	26.1
	.005	43.1	42.8	42.6	42.5	42.3	42.2	42.1	42.0	41.8
	.001	127	126	126	125	125	125	124	124	124
4	.75	.664	.683	.692	.702	.712	.718	.722	.733	.743
	.50	1.14	1.15	1.16	1.16	1.17	1.18	1.18	1.18	1.19
	.25	2.08	2.08	2.08	2.08	2.08	2.08	2.08	2.08	2.08
	.10	3.87	3.84	3.83	3.82	3.80	3.79	3.79	3.78	3.76
	.05	5.86	5.80	5.77	5.75	5.72	5.70	5.69	5.66	5.63
	.025	8.66	8.56	8.51	8.46	8.41	8.38	8.36	8.31	8.26
	.01	14.2	14.0	13.9	13.8	13.7	13.7	13.7	13.6	13.5
	.005	20.4	20.2	20.0	19.9	19.8	19.7	19.6	19.5	19.3
	.001	46.8	46.1	45.8	45.4	45.1	44.9	44.8	44.4	44.0
5	.75	.669	.690	.700	.711	.722	.728	.732	.743	.755
	.50	1.10	1.11	1.12	1.12	1.13	1.14	1.14	1.14	1.15
	.25	1.89	1.88	1.88	1.88	1.88	1.87	1.87	1.87	1.87
	.10	3.24	3.21	3.19	3.17	3.16	3.15	3.14	3.12	3.10
	.05	4.62	4.56	4.53	4.50	4.46	4.44	4.43	4.40	4.36
	.025	6.43	6.33	6.28	6.23	6.18	6.14	6.12	6.07	6.02
	.01	9.72	9.55	9.47	9.38	9.29	9.24	9.20	9.11	9.02
	.005	13.1	12.9	12.8	12.7	12.5	12.4	12.4	12.3	12.1
	.001	25.9	25.4	25.1	24.9	24.6	24.4	24.3	24.1	23.7

$$\nu_1$$

(degrees of freedom of numerator mean squares)

15	20	24	30	40	50	60	120	∞	α	
.675	.696	.707	.718	.729	.736	.741	.753	.765	.75	
1.07	1.08	1.09	1.10	1.10	1.11	1.11	1.12	1.12	.50	
1.76	1.76	1.75	1.75	1.75	1.74	1.74	1.74	1.74	.25	
2.87	2.84	2.82	2.80	2.78	2.77	2.76	2.74	2.72	.10	
3.94	3.87	3.84	3.81	3.77	3.75	3.74	3.70	3.67	.05	**6**
5.27	5.17	5.12	5.07	5.01	4.98	4.96	4.90	4.85	.025	
7.56	7.40	7.31	7.23	7.14	7.09	7.06	6.97	6.88	.01	
9.81	9.59	9.47	9.36	9.24	9.17	9.12	9.00	8.88	.005	
17.6	17.1	16.9	16.7	16.4	16.3	16.2	16.0	15.8	.001	
.679	.702	.713	.725	.737	.745	.749	.762	.775	.75	
1.05	1.07	1.07	1.08	1.08	1.09	1.09	1.10	1.10	.50	
1.68	1.67	1.67	1.66	1.66	1.65	1.65	1.65	1.65	.25	
2.63	2.59	2.58	2.56	2.54	2.52	2.51	2.49	2.47	.10	
3.51	3.44	3.41	3.38	3.34	3.32	3.30	3.27	3.23	.05	**7**
4.57	4.47	4.42	4.36	4.31	4.27	4.25	4.20	4.14	.025	
6.31	6.16	6.07	5.99	5.91	5.86	5.82	5.74	5.65	.01	
7.97	7.75	7.65	7.53	7.42	7.35	7.31	7.19	7.08	.005	
13.3	12.9	12.7	12.5	12.3	12.2	12.1	11.9	11.7	.001	
.684	.707	.718	.730	.743	.751	.756	.769	.783	.75	
1.04	1.05	1.06	1.07	1.08	1.07	1.08	1.08	1.09	.50	
1.62	1.61	1.60	1.60	1.59	1.59	1.59	1.58	1.58	.25	
2.46	2.42	2.40	2.38	2.36	2.35	2.34	2.32	2.29	.10	
3.22	3.15	3.12	3.08	3.04	3.02	3.01	2.97	2.93	.05	**8**
4.10	4.00	3.95	3.89	3.84	3.80	3.78	3.73	3.67	.025	
5.52	5.36	5.28	5.20	5.12	5.07	5.03	4.95	4.86	.01	
6.81	6.61	6.50	6.40	6.29	6.22	6.18	6.06	5.95	.005	
10.8	10.5	10.3	10.1	9.9	9.8	9.7	9.5	9.3	.001	
.687	.711	.723	.736	.749	.757	.762	.776	.791	.75	
1.03	1.04	1.05	1.05	1.06	1.07	1.07	1.07	1.08	.50	
1.57	1.56	1.56	1.55	1.55	1.54	1.54	1.53	1.53	.25	
2.34	2.30	2.28	2.25	2.23	2.22	2.21	2.18	2.16	.10	
3.01	2.94	2.90	2.86	2.83	2.81	2.79	2.75	2.71	.05	**9**
3.77	3.67	3.61	3.56	3.51	3.47	3.45	3.39	3.33	.025	
4.96	4.81	4.73	4.65	4.57	4.52	4.48	4.40	4.31	.01	
6.03	5.83	5.73	5.62	5.52	5.45	5.41	5.30	5.19	.005	
9.24	8.90	8.72	8.55	8.37	8.26	8.19	8.00	7.81	.001	
.691	.714	.727	.740	.754	.762	.767	.782	.797	.75	
1.02	1.03	1.04	1.05	1.05	1.06	1.06	1.06	1.07	.50	
1.53	1.52	1.52	1.51	1.51	1.50	1.50	1.49	1.48	.25	
2.24	2.20	2.18	2.16	2.13	2.12	2.11	2.08	2.06	.10	
2.85	2.77	2.74	2.70	2.66	2.64	2.62	2.58	2.54	.05	**10**
3.52	3.42	3.37	3.31	3.26	3.22	3.20	3.14	3.08	.025	
4.56	4.41	4.33	4.25	4.17	4.12	4.08	4.00	3.91	.01	
5.47	5.27	5.17	5.07	4.97	4.90	4.86	4.75	4.64	.005	
8.13	7.80	7.64	7.47	7.30	7.19	7.12	6.94	6.76	.001	

ν_2 (degrees of freedom of denominator mean squares)

TABLE F Critical values of the F-distribution (*continued*)

ν_1

(degrees of freedom of numerator mean squares)

	α	15	20	24	30	40	50	60	120	∞
	.75	.694	.719	.730	.744	.758	.767	.773	.788	.803
	.50	1.02	1.03	1.03	1.04	1.05	1.05	1.05	1.06	1.06
	.25	1.50	1.49	1.49	1.48	1.47	1.47	1.47	1.46	1.45
	.10	2.17	2.12	2.10	2.08	2.05	2.04	2.03	2.00	1.97
11	.05	2.72	2.65	2.61	2.57	2.53	2.51	2.49	2.45	2.40
	.025	3.33	3.23	3.17	3.12	3.06	3.02	3.00	2.94	2.88
	.01	4.25	4.10	4.02	3.94	3.86	3.81	3.78	3.69	3.60
	.005	5.05	4.86	4.76	4.65	4.55	4.49	4.45	4.34	4.23
	.001	7.32	7.01	6.85	6.68	6.52	6.42	6.35	6.17	6.00
	.75	.695	.721	.734	.748	.762	.771	.777	.792	.808
	.50	1.01	1.02	1.03	1.03	1.04	1.04	1.05	1.05	1.06
	.25	1.48	1.47	1.46	1.45	1.45	1.44	1.44	1.43	1.42
	.10	2.11	2.06	2.04	2.01	1.99	1.97	1.96	1.93	1.90
12	.05	2.62	2.54	2.51	2.47	2.43	2.40	2.38	2.34	2.30
	.025	3.18	3.07	3.02	2.96	2.91	2.87	2.85	2.79	2.72
	.01	4.01	3.86	3.78	3.70	3.62	3.57	3.54	3.45	3.36
	.005	4.72	4.53	4.43	4.33	4.23	4.16	4.12	4.01	3.90
	.001	6.71	6.40	6.25	6.09	5.93	5.83	5.76	5.59	5.42
	.75	.697	.723	.737	.751	.766	.775	.781	.797	.813
	.50	1.01	1.02	1.02	1.03	1.04	1.04	1.04	1.05	1.05
	.25	1.46	1.45	1.44	1.43	1.42	1.42	1.42	1.41	1.40
	.10	2.05	2.01	1.98	1.96	1.93	1.92	1.90	1.88	1.85
13	.05	2.53	2.46	2.42	2.38	2.34	2.31	2.30	2.25	2.21
	.025	3.05	2.95	2.89	2.84	2.78	2.74	2.72	2.66	2.60
	.01	3.82	3.66	3.59	3.51	3.43	3.37	3.34	3.25	3.17
	.005	4.46	4.27	4.17	4.07	3.97	3.91	3.87	3.76	2.65
	.001	6.23	5.93	5.78	5.63	5.47	5.37	5.30	5.14	4.97
	.75	.699	.726	.740	.754	.769	.778	.785	.801	.818
	.50	1.00	1.01	1.02	1.03	1.03	1.04	1.04	1.04	1.05
	.25	1.44	1.43	1.42	1.41	1.41	1.40	1.40	1.39	1.38
	.10	2.01	1.96	1.94	1.91	1.89	1.87	1.86	1.83	1.80
14	.05	2.46	2.39	2.35	2.31	2.27	2.24	2.22	2.18	2.13
	.025	2.95	2.84	2.79	2.73	2.67	2.64	2.61	2.55	2.49
	.01	3.66	3.51	3.43	3.35	3.27	3.21	3.18	3.09	3.00
	.005	4.25	4.06	3.96	3.86	3.76	3.70	3.66	3.55	3.44
	.001	5.85	5.56	5.41	5.25	5.10	5.00	4.94	4.77	4.60
	.75	.701	.728	.742	.757	.772	.782	.788	.805	.822
	.50	1.00	1.01	1.02	1.02	1.03	1.03	1.03	1.04	1.05
	.25	1.43	1.41	1.41	1.40	1.39	1.38	1.38	1.37	1.36
	.10	1.97	1.92	1.90	1.87	1.85	1.83	1.82	1.79	1.76
15	.05	2.40	2.33	2.29	2.25	2.20	2.18	2.16	2.11	2.07
	.025	2.86	2.76	2.70	2.64	2.59	2.55	2.52	2.46	2.40
	.01	3.52	3.37	3.29	3.21	3.13	3.08	3.05	2.96	2.87
	.005	4.07	3.88	3.79	3.69	3.59	3.52	3.48	3.37	3.26
	.001	5.54	5.25	5.10	4.95	4.80	4.70	4.64	4.47	4.31

ν_2 (degrees of freedom of denominator mean squares)

$$v_1$$
(degrees of freedom of numerator mean squares)

15	20	24	30	40	50	60	120	∞	α	
.703	.730	.744	.759	.775	.785	.791	.808	.826	.75	
.997	1.01	1.01	1.02	1.03	1.03	1.03	1.04	1.04	.50	
1.41	1.40	1.39	1.38	1.37	1.37	1.36	1.35	1.34	.25	
1.94	1.89	1.87	1.84	1.81	1.79	1.78	1.75	1.72	.10	
2.35	2.28	2.24	2.19	2.15	2.12	2.11	2.06	2.01	.05	**16**
2.79	2.68	2.63	2.57	2.51	2.47	2.45	2.38	2.32	.025	
3.41	3.26	3.18	3.10	3.02	2.97	2.93	2.84	2.75	.01	
3.92	3.73	3.64	3.54	3.44	3.37	3.33	3.22	3.11	.005	
5.27	4.99	4.85	4.70	4.54	4.45	4.39	4.23	4.06	.001	
.704	.732	.746	.762	.777	.787	.794	.811	.830	.75	
.995	1.01	1.01	1.02	1.02	1.03	1.03	1.03	1.04	.50	
1.40	1.39	1.38	1.37	1.36	1.36	1.35	1.34	1.33	.25	
1.91	1.86	1.84	1.81	1.78	1.76	1.75	1.72	1.69	.10	
2.31	2.23	2.19	2.15	2.10	2.08	2.06	2.01	1.96	.05	**17**
2.72	2.62	2.56	2.50	2.44	2.41	2.38	2.32	2.25	.025	
3.31	3.16	3.08	3.00	2.92	2.87	2.83	2.75	2.65	.01	
3.79	3.61	3.51	3.41	3.31	3.25	3.21	3.10	2.98	.005	
5.05	4.78	4.63	4.48	4.33	4.24	4.18	4.02	3.85	.001	
.706	.733	.748	.764	.780	.790	.797	.814	.833	.75	
.992	1.00	1.01	1.02	1.02	1.02	1.03	1.03	1.04	.50	
1.39	1.38	1.37	1.36	1.35	1.34	1.34	1.33	1.32	.25	
1.89	1.84	1.81	1.78	1.75	1.74	1.72	1.69	1.66	.10	
2.27	2.19	2.15	2.11	2.06	2.04	2.02	1.97	1.92	.05	**18**
2.67	2.56	2.50	2.44	2.38	2.35	2.32	2.26	2.19	.025	
3.23	3.08	3.00	2.92	2.84	2.78	2.75	2.66	2.57	.01	
3.68	3.50	3.40	3.30	3.20	3.14	3.10	2.99	2.87	.005	
4.87	4.59	4.45	4.30	4.15	4.06	4.00	3.84	3.67	.001	
.707	.735	.750	.766	.782	.792	.799	.817	.836	.75	
.990	1.00	1.01	1.01	1.02	1.02	1.02	1.03	1.04	.50	
1.38	1.37	1.36	1.35	1.34	1.33	1.33	1.32	1.30	.25	
1.86	1.81	1.79	1.76	1.73	1.71	1.70	1.67	1.63	.10	
2.23	2.16	2.11	2.07	2.03	2.00	1.98	1.93	1.88	.05	**19**
2.62	2.51	2.45	2.39	2.33	2.30	2.27	2.20	2.13	.025	
3.15	3.00	2.92	2.84	2.76	2.71	2.67	2.58	2.49	.01	
3.59	3.40	3.31	3.21	3.11	3.04	3.00	2.89	2.78	.005	
4.70	4.43	4.29	4.14	3.99	3.90	3.84	3.68	3.51	.001	
.708	.736	.751	.767	.784	.794	.801	.820	.840	.75	
.989	1.00	1.01	1.01	1.02	1.02	1.02	1.03	1.03	.50	
1.37	1.36	1.35	1.34	1.33	1.33	1.32	1.31	1.29	.25	
1.84	1.79	1.77	1.74	1.71	1.69	1.68	1.64	1.61	.10	
2.20	2.12	2.08	2.04	1.99	1.97	1.95	1.90	1.84	.05	**20**
2.57	2.46	2.41	2.35	2.29	2.25	2.22	2.16	2.09	.025	
3.09	2.94	2.86	2.78	2.69	2.64	2.61	2.52	2.42	.01	
3.50	3.32	3.22	3.12	3.02	2.96	2.92	2.81	2.69	.005	
4.56	4.29	4.15	4.00	3.86	3.76	3.70	3.54	3.38	.001	

v_2 (degrees of freedom of denominator mean squares)

TABLE F   Critical values of the *F*-distribution (*continued*)

ν_1

(degrees of freedom of numerator mean squares)

	α	15	20	24	30	40	50	60	120	∞
	.75	.709	.738	.753	.769	.786	.796	.803	.822	.842
	.50	.987	.998	1.00	1.01	1.02	1.02	1.02	1.03	1.03
	.25	1.37	1.35	1.34	1.33	1.32	1.32	1.31	1.30	1.28
	.10	1.83	1.78	1.75	1.72	1.69	1.67	1.66	1.62	1.59
21	.05	2.18	2.10	2.05	2.01	1.96	1.94	1.92	1.87	1.81
	.025	2.53	2.42	2.37	2.31	2.25	2.21	2.18	2.11	2.04
	.01	3.03	2.88	2.80	2.72	2.64	2.58	2.55	2.46	2.36
	.005	3.43	3.24	3.15	3.05	2.95	2.88	2.84	2.73	2.61
	.001	4.44	4.17	4.03	3.88	3.74	3.64	3.58	3.42	3.26
	.75	.710	.739	.754	.770	.787	.798	.805	.824	.845
	.50	.986	.997	1.00	1.01	1.01	1.02	1.02	1.03	1.03
	.25	1.36	1.34	1.33	1.32	1.31	1.31	1.30	1.29	1.28
	.10	1.81	1.76	1.73	1.70	1.67	1.65	1.64	1.60	1.57
22	.05	2.15	2.07	2.03	1.98	1.94	1.91	1.89	1.84	1.78
	.025	2.50	2.39	2.33	2.27	2.21	2.17	2.14	2.08	2.00
	.01	2.98	2.83	2.75	2.67	2.58	2.53	2.50	2.40	2.31
	.005	3.36	3.18	3.08	2.98	2.88	2.82	2.77	2.66	2.55
	.001	4.33	4.06	3.92	3.78	3.63	3.54	3.48	3.32	3.15
	.75	.711	.740	.756	.772	.789	.800	.807	.827	.847
	.50	.984	.996	1.00	1.01	1.01	1.02	1.02	1.02	1.03
	.25	1.35	1.34	1.33	1.32	1.31	1.30	1.30	1.28	1.27
	.10	1.80	1.74	1.72	1.69	1.66	1.64	1.62	1.59	1.55
23	.05	2.13	2.05	2.01	1.96	1.91	1.89	1.86	1.81	1.76
	.025	2.47	2.36	2.30	2.24	2.18	2.14	2.11	2.04	1.97
	.01	2.93	2.78	2.70	2.62	2.54	2.48	2.45	2.35	2.26
	.005	3.30	3.12	3.02	2.92	2.82	2.76	2.71	2.60	2.48
	.001	4.23	3.96	3.82	3.68	3.53	3.44	3.38	3.22	3.05
	.75	.712	.741	.757	.773	.791	.802	.809	.829	.850
	.50	.983	.994	1.00	1.01	1.01	1.02	1.02	1.02	1.03
	.25	1.35	1.33	1.32	1.31	1.30	1.29	1.29	1.28	1.26
	.10	1.78	1.73	1.70	1.67	1.64	1.62	1.61	1.57	1.53
24	.05	2.11	2.03	1.98	1.94	1.89	1.86	1.84	1.79	1.73
	.025	2.44	2.33	2.27	2.21	2.15	2.11	2.08	2.01	1.94
	.01	2.89	2.74	2.66	2.58	2.49	2.44	2.40	2.31	2.21
	.005	3.25	3.06	2.97	2.87	2.77	2.70	2.66	2.55	2.43
	.001	4.14	3.87	3.74	3.59	3.45	3.36	3.29	3.14	2.97
	.75	.712	.742	.758	.775	.792	.803	.811	.831	.852
	.50	.982	.993	.999	1.00	1.01	1.01	1.02	1.02	1.03
	.25	1.34	1.33	1.32	1.31	1.29	1.29	1.28	1.27	1.25
	.10	1.77	1.72	1.69	1.66	1.63	1.61	1.59	1.56	1.52
25	.05	2.09	2.01	1.96	1.92	1.87	1.84	1.82	1.77	1.71
	.025	2.41	2.30	2.24	2.18	2.12	2.08	2.05	1.98	1.91
	.01	2.85	2.70	2.62	2.54	2.45	2.40	2.36	2.27	2.17
	.005	3.20	3.01	2.92	2.82	2.72	2.65	2.61	2.50	2.38
	.001	4.06	3.97	3.66	3.52	3.37	3.28	3.22	3.06	2.89

ν_2 (degrees of freedom of denominator mean squares)

$$\nu_1$$

(degrees of freedom of numerator mean squares)

15	20	24	30	40	50	60	120	∞	α	
.713	.743	.759	.776	.793	.805	.812	.832	.854	.75	
.981	.992	.998	1.00	1.01	1.01	1.01	1.02	1.03	.50	
1.34	1.32	1.31	1.30	1.29	1.28	1.28	1.26	1.25	.25	
1.76	1.71	1.68	1.65	1.61	1.59	1.58	1.54	1.50	.10	
2.07	1.99	1.95	1.90	1.85	1.82	1.80	1.75	1.69	.05	**26**
2.39	2.28	2.22	2.16	2.09	2.05	2.03	1.95	1.88	.025	
2.81	2.66	2.58	2.50	2.42	2.36	2.33	2.23	2.13	.01	
3.15	2.97	2.87	2.77	2.67	2.61	2.56	2.45	2.33	.005	
3.99	3.72	3.59	3.44	3.30	3.21	3.15	2.99	2.82	.001	
.714	.744	.760	.777	.795	.806	.814	.834	.856	.75	
.980	.991	.997	1.00	1.01	1.01	1.01	1.02	1.03	.50	
1.33	1.32	1.31	1.30	1.28	1.28	1.27	1.26	1.24	.25	
1.75	1.70	1.67	1.64	1.60	1.58	1.57	1.53	1.49	.10	
2.06	1.97	1.93	1.88	1.84	1.81	1.79	1.73	1.67	.05	**27**
2.36	2.25	2.19	2.13	2.07	2.03	2.00	1.93	1.85	.025	
2.78	2.63	2.55	2.47	2.38	2.33	2.29	2.20	2.10	.01	
3.11	2.93	2.83	2.73	2.63	2.57	2.52	2.41	2.29	.005	
3.92	3.66	3.52	3.38	3.23	3.14	3.08	2.92	2.75	.001	
.714	.745	.761	.778	.796	.807	.815	.856	.858	.75	
.979	.990	.996	1.00	1.01	1.01	1.01	1.02	1.02	.50	
1.33	1.31	1.30	1.29	1.28	1.27	1.27	1.25	1.24	.25	
1.74	1.69	1.66	1.63	1.59	1.57	1.56	1.52	1.48	.10	
2.04	1.96	1.91	1.87	1.82	1.79	1.77	1.71	1.65	.05	**28**
2.34	2.23	2.17	2.11	2.05	2.01	1.98	1.91	1.83	.025	
2.75	2.60	2.52	2.44	2.35	2.30	2.26	2.17	2.06	.01	
3.07	2.89	2.79	2.69	2.59	2.53	2.48	2.37	2.25	.005	
3.86	3.60	3.46	3.32	3.18	3.09	3.02	2.86	2.69	.001	
.715	.745	.762	.779	.797	.809	.816	.837	.860	.75	
.979	.990	.996	1.00	1.01	1.01	1.01	1.02	1.02	.50	
1.32	1.31	1.30	1.29	1.27	1.27	1.26	1.25	1.23	.25	
1.73	1.68	1.65	1.62	1.58	1.56	1.55	1.51	1.47	.10	
2.03	1.94	1.90	1.85	1.81	1.78	1.75	1.70	1.64	.05	**29**
2.32	2.21	2.15	2.09	2.03	1.99	1.96	1.89	1.81	.025	
2.73	2.57	2.49	2.41	2.33	2.27	2.23	2.14	2.03	.01	
3.04	2.86	2.76	2.66	2.56	2.49	2.45	2.33	2.21	.005	
3.80	3.54	3.41	3.27	3.12	3.03	2.97	2.81	2.64	.001	
.716	.746	.763	.780	.798	.810	.818	.839	.862	.75	
.978	.989	.994	1.00	1.01	1.01	1.01	1.02	1.02	.50	
1.32	1.30	1.29	1.28	1.27	1.26	1.26	1.24	1.23	.25	
1.72	1.67	1.64	1.61	1.57	1.55	1.54	1.50	1.46	.10	
2.01	1.93	1.89	1.84	1.79	1.76	1.74	1.68	1.62	.05	**30**
2.31	2.20	2.14	2.07	2.01	1.97	1.94	1.87	1.79	.025	
2.70	2.55	2.47	2.39	2.30	2.25	2.21	2.11	2.01	.01	
3.01	2.82	2.73	2.63	2.52	2.46	2.42	2.30	2.18	.005	
3.75	3.49	3.36	3.22	3.07	2.98	2.92	2.76	2.59	.001	

ν_2 (degrees of freedom of denominator mean squares)

TABLE F Critical values of the F-distribution (*continued*)

ν_1

(degrees of freedom of numerator mean squares)

ν_2	α	15	20	24	30	40	50	60	120	∞
40	.75	.720	.752	.769	.787	.806	.819	.828	.851	.877
	.50	.972	.983	.989	.994	1.00	1.00	1.01	1.01	1.02
	.25	1.30	1.28	1.26	1.25	1.24	1.23	1.22	1.21	1.19
	.10	1.66	1.61	1.57	1.54	1.51	1.48	1.47	1.42	1.38
	.05	1.92	1.84	1.79	1.74	1.69	1.66	1.64	1.58	1.51
	.025	2.18	2.07	2.01	1.94	1.88	1.83	1.80	1.72	1.64
	.01	2.52	2.37	2.29	2.20	2.11	2.06	2.02	1.92	1.80
	.005	2.78	2.60	2.50	2.40	2.30	2.23	2.18	2.06	1.93
	.001	3.40	3.15	3.01	2.87	2.73	2.64	2.57	2.41	2.23
60	.75	.725	.758	.776	.796	.816	.830	.840	.865	.896
	.50	.967	.978	.983	.989	.994	.998	1.00	1.01	1.01
	.25	1.27	1.25	1.24	1.22	1.21	1.20	1.19	1.17	1.15
	.10	1.60	1.54	1.51	1.48	1.44	1.41	1.40	1.35	1.29
	.05	1.84	1.75	1.70	1.65	1.59	1.56	1.53	1.47	1.39
	.025	2.06	1.94	1.88	1.82	1.74	1.70	1.67	1.58	1.48
	.01	2.35	2.20	2.12	2.03	1.94	1.88	1.84	1.73	1.60
	.005	2.57	2.39	2.29	2.19	2.08	2.01	1.96	1.83	1.69
	.001	3.08	2.83	2.69	2.55	2.41	2.32	2.25	2.08	1.89
120	.75	.730	.765	.784	.805	.828	.843	.853	.884	.923
	.50	.961	.972	.978	.983	.989	.992	.994	1.00	1.01
	.25	1.24	1.22	1.21	1.19	1.18	1.17	1.16	1.13	1.10
	.10	1.55	1.48	1.45	1.41	1.37	1.34	1.32	1.26	1.19
	.05	1.75	1.66	1.61	1.55	1.50	1.46	1.43	1.35	1.25
	.025	1.95	1.82	1.76	1.69	1.61	1.56	1.53	1.43	1.31
	.01	2.19	2.03	1.95	1.86	1.76	1.70	1.66	1.53	1.38
	.005	2.37	2.19	2.09	1.98	1.87	1.80	1.75	1.61	1.43
	.001	2.78	2.53	2.40	2.26	2.11	2.02	1.95	1.76	1.54
∞	.75	.736	.773	.793	.816	.842	.860	.872	.910	1.00
	.50	.956	.967	.972	.978	.983	.987	.989	.994	1.00
	.25	1.22	1.19	1.18	1.16	1.14	1.13	1.12	1.08	1.00
	.10	1.49	1.42	1.38	1.34	1.30	1.26	1.24	1.17	1.00
	.05	1.67	1.57	1.52	1.46	1.39	1.35	1.32	1.22	1.00
	.025	1.83	1.71	1.64	1.57	1.48	1.43	1.39	1.27	1.00
	.01	2.04	1.88	1.79	1.70	1.59	1.52	1.47	1.32	1.00
	.005	2.19	2.00	1.90	1.79	1.67	1.59	1.53	1.36	1.00
	.001	2.51	2.27	2.13	1.99	1.84	1.73	1.66	1.45	1.00

ν_2 (degrees of freedom of denominator mean squares)

TABLE **G**   Critical values of F_{max}

This table furnishes critical values for the maximum F-ratio distribution. The maximum observed F-ratio is computed as s_{max}^2/s_{min}^2 and compared with the critical value for a samples and degrees of freedom $v = n - 1$, where n is the sample size. The critical values of F_{max} are tabulated for number of samples a from 2 to 12 in increments of one and for degrees of freedom v from 2 to 10 in increments of one, as well as for $v = 12, 15, 20, 30, 60$, and ∞. Corresponding to each value of a and v are two critical values of F_{max} representing the upper 5% and 1% points. The corresponding probabilities $\alpha = 0.05$ and 0.01 represent *one tail* of the F_{max}-distribution. The critical values are given to a varying number of decimal places, depending on the exactness of the computation. Values for $a = 2$ and $v = 2$ and ∞ are exact. In other places in the table the third digit may be slightly in error. The third digit for $v = 3$ (in parentheses) is the most uncertain value in the table.

To find the critical values of F_{max} in an example with 6 samples and 10 items per sample, we look up $a = 6$ and $v = n - 1 = 9$ to obtain $F_{max} = 7.80$ and 12.1 at the 5% and 1% levels, respectively.

This table is used for testing homogeneity of variances (Section 13.3).

This table was copied from H. A. David (*Biometrika* **39**:422–424, 1952) with the permission of the publisher.

TABLE G   Critical values of F_{max}

v \ a	2	3	4	5	6	7	8	9	10	11	12
2	39.0	87.5	142.	202.	266.	333.	403.	475.	550.	626.	704.
	199.	448.	729.	1036.	1362.	1705.	2063.	2432.	2813.	3204.	3605.
3	15.4	27.8	39.2	50.7	62.0	72.9	83.5	93.9	104.	114.	124.
	47.5	85.	120.	151.	184.	21(6)	24(9)	28(1)	31(0)	33(7)	36(1)
4	9.60	15.5	20.6	25.2	29.5	33.6	37.5	41.1	44.6	48.0	51.4
	23.2	37.	49.	59.	69.	79.	89.	97.	106.	113.	120.
5	7.15	10.8	13.7	16.3	18.7	20.8	22.9	24.7	26.5	28.2	29.9
	14.9	22.	28.	33.	38.	42.	46.	50.	54.	57.	60.
6	5.82	8.38	10.4	12.1	13.7	15.0	16.3	17.5	18.6	19.7	20.7
	11.1	15.5	19.1	22.	25.	27.	30.	32.	34.	36.	37.
7	4.99	6.94	8.44	9.70	10.8	11.8	12.7	13.5	14.3	15.1	15.8
	8.89	12.1	14.5	16.5	18.4	20.	22.	23.	24.	26.	27.
8	4.43	6.00	7.18	8.12	9.03	9.78	10.5	11.1	11.7	12.2	12.7
	7.50	9.9	11.7	13.2	14.5	15.8	16.9	17.9	18.9	19.8	21.
9	4.03	5.34	6.31	7.11	7.80	8.41	8.95	9.45	9.91	10.3	10.7
	6.54	8.5	9.9	11.1	12.1	13.1	13.9	14.7	15.3	16.0	16.6
10	3.72	4.85	5.67	6.34	6.92	7.42	7.87	8.28	8.66	9.01	9.34
	5.85	7.4	8.6	9.6	10.4	11.1	11.8	12.4	12.9	13.4	13.9
12	3.28	4.16	4.79	5.30	5.72	6.09	6.42	6.72	7.00	7.25	7.48
	4.91	6.1	6.9	7.6	8.2	8.7	9.1	9.5	9.9	10.2	10.6
15	2.86	3.54	4.01	4.37	4.68	4.95	5.19	5.40	5.59	5.77	5.93
	4.07	4.9	5.5	6.0	6.4	6.7	7.1	7.3	7.5	7.8	8.0
20	2.46	2.95	3.29	3.54	3.76	3.94	4.10	4.24	4.37	4.49	4.59
	3.32	3.8	4.3	4.6	4.9	5.1	5.3	5.5	5.6	5.8	5.9
30	2.07	2.40	2.61	2.78	2.91	3.02	3.12	3.21	3.29	3.36	3.39
	2.63	3.0	3.3	3.4	3.6	3.7	3.8	3.9	4.0	4.1	4.2
60	1.67	1.85	1.96	2.04	2.11	2.17	2.22	2.26	2.30	2.33	2.36
	1.96	2.2	2.3	2.4	2.4	2.5	2.5	2.6	2.6	2.7	2.7
∞	1.00	1.00	1.00	1.00	1.00	1.00	1.00	1.00	1.00	1.00	1.00
	1.00	1.00	1.00	1.00	1.00	1.00	1.00	1.00	1.00	1.00	1.00

TABLE **H** Rankits (normal order statistics)

This table lists the expected values of ranked normal deviates, for each position in the rank order, of samples ranging in size from $n = 2$ to $n = 50$. These values, also known as rankits, can be thought to have been obtained as follows. If one samples n items repeatedly from a standard normal population ($\mu = 0$, $\sigma = 1$, items normally distributed) and ranks the items in each sample by order of magnitude, the first rankit value is the mean of the first item from every sample. The second rankit is the mean of the second item from every sample, and so forth. The rankits, in population standard deviation units, are given to three significant decimal places for sample sizes up to $n = 20$, and to two decimal places beyond that number.

Rankits are symmetrical about their median. The table shows only the rankits greater than their median. For odd-numbered samples the median rankit is zero and is also shown. For example, the rankits for sample size $n = 3$ are (from the left to the right tail of the distribution) -0.846, 0.000, 0.846, and for $n = 4$ they are -1.029, -0.297, 0.297, 1.029. Thus, the second largest item in a normally distributed sample of 4 items is, on the average, 0.297 standard deviations above the mean. In a sample of 40 items, however, the second largest item is 1.75 standard deviations above the mean.

Rankits are used in the graphic analysis of small samples (tests for normality and estimation of mean and standard deviation; Section 6.7 and Box 6.3).

Values in this table for sample sizes $n = 2$ to $n = 20$ were taken from table 28 in E. S. Pearson and H. O. Hartley, *Biometrika Tables for Statisticians,* Vol. 1 (Cambridge University Press, 1958) with permission of the publishers. The values between $n = 21$ and $n = 50$ are reproduced form table XX in R. A. Fisher and F. Yates, *Statistical Tables for Biological, Agricultural and Medical Research,* 5th ed. (Oliver & Boyd, Edinburgh, 1958) with permission of the authors and their publisher.

TABLE **H**   Rankits (normal order statistics)

Rank order \ n	2	3	4	5	6	7	8	9	10
1	0.564	0.846	1.029	1.163	1.267	1.352	1.424	1.485	1.539
2		.000	.297	.495	.642	.757	.852	.932	1.001
3				.000	.202	.353	.473	.572	.656
4						.000	.153	.275	.376
5								.000	.123

Rank order \ n	11	12	13	14	15	16	17	18	19	20
1	1.586	1.629	1.668	1.703	1.736	1.766	1.794	1.820	1.844	1.867
2	1.062	1.116	1.164	1.208	1.248	1.285	1.319	1.350	1.380	1.408
3	.729	.793	.850	.901	.948	.990	1.029	1.066	1.099	1.131
4	.462	.537	.603	.662	.715	.763	.807	.848	.886	.921
5	.225	.312	.388	.456	.516	.570	.619	.665	.707	.745
6	.000	.103	.190	.267	.335	.396	.451	.502	.548	.590
7		.000	.088	.165	.234	.295	.351	.402	.448	
8				.000	.077	.146	.208	.264	.315	
9					.000	.069	.131	.187		
10							.000	.062		

Rank order \ n	21	22	23	24	25	26	27	28	29	30
1	1.89	1.91	1.93	1.95	1.97	1.98	2.00	2.01	2.03	2.04
2	1.43	1.46	1.48	1.50	1.52	1.54	1.56	1.58	1.60	1.62
3	1.16	1.19	1.21	1.24	1.26	1.29	1.31	1.33	1.35	1.36
4	.95	.98	1.01	1.04	1.07	1.09	1.11	1.14	1.16	1.18
5	.78	.82	.85	.88	.91	.93	.96	.98	1.00	1.03
6	.63	.67	.70	.73	.76	.79	.82	.85	.87	.89
7	.49	.53	.57	.60	.64	.67	.70	.73	.75	.78
8	.36	.41	.45	.48	.52	.55	.58	.61	.64	.67
9	.24	.29	.33	.37	.41	.44	.48	.51	.54	.57
10	.12	.17	.22	.26	.30	.34	.38	.41	.44	.47
11	.00	.06	.11	.16	.20	.24	.28	.32	.35	.38
12		.00	.05	.10	.14	.19	.22	.26	.29	
13			.00	.05	.09	.13	.17	.21		
14				.00	.04	.09	.12			
15					.00	.04				

TABLE H Rankits (normal order statistics) (*continued*)

Rank order	*n* 31	32	33	34	35	36	37	38	39	40
1	2.06	2.07	2.08	2.09	2.11	2.12	2.13	2.14	2.15	2.16
2	1.63	1.65	1.66	1.68	1.69	1.70	1.72	1.73	1.74	1.75
3	1.38	1.40	1.42	1.43	1.45	1.46	1.48	1.49	1.50	1.52
4	1.20	1.22	1.23	1.25	1.27	1.28	1.30	1.32	1.33	1.34
5	1.05	1.07	1.09	1.11	1.12	1.14	1.16	1.17	1.19	1.20
6	.92	.94	.96	.98	1.00	1.02	1.03	1.05	1.07	1.08
7	.80	.82	.85	.87	.89	.91	.92	.94	.96	.98
8	.69	.72	.74	.76	.79	.81	.83	.85	.86	.88
9	.60	.62	.65	.67	.69	.72	.73	.75	.77	.79
10	.50	.53	.56	.58	.60	.63	.65	.67	.69	.71
11	.41	.44	.47	.50	.52	.54	.57	.59	.61	.63
12	.33	.36	.39	.41	.44	.47	.49	.51	.54	.56
13	.24	.28	.31	.34	.36	.39	.42	.44	.46	.49
14	.16	.20	.23	.26	.29	.32	.34	.37	.39	.42
15	.08	.12	.15	.18	.22	.24	.27	.30	.33	.35
16	.00	.04	.08	.11	.14	.17	.20	.23	.26	.28
17			.00	.04	.07	.10	.14	.16	.19	.22
18					.00	.03	.07	.10	.13	.16
19							.00	.03	.06	.09
20									.00	.03

Rank order	*n* 41	42	43	44	45	46	47	48	49	50
1	2.17	2.18	2.19	2.20	2.21	2.22	2.22	2.23	2.24	2.25
2	1.76	1.78	1.79	1.80	1.81	1.82	1.83	1.84	1.85	1.85
3	1.53	1.54	1.55	1.57	1.58	1.59	1.60	1.61	1.62	1.63
4	1.36	1.37	1.38	1.40	1.41	1.42	1.43	1.44	1.45	1.46
5	1.22	1.23	1.25	1.26	1.27	1.28	1.30	1.31	1.32	1.33
6	1.10	1.11	1.13	1.14	1.16	1.17	1.18	1.19	1.21	1.22
7	.99	1.01	1.02	1.04	1.05	1.07	1.08	1.09	1.11	1.12
8	.90	.91	.93	.95	.96	.98	.99	1.00	1.02	1.03
9	.81	.83	.84	.86	.88	.89	.91	.92	.94	.95
10	.73	.75	.76	.78	.80	.81	.83	.84	.86	.87
11	.65	.67	.69	.71	.72	.74	.76	.77	.79	.80
12	.58	.60	.62	.64	.65	.67	.69	.70	.72	.74
13	.51	.53	.55	.57	.59	.60	.62	.64	.66	.67
14	.44	.46	.48	.50	.52	.54	.56	.58	.59	.61
15	.37	.40	.42	.44	.46	.48	.50	.52	.53	.55
16	.31	.33	.36	.38	.40	.42	.44	.46	.48	.49
17	.25	.27	.29	.32	.34	.36	.38	.40	.42	.44
18	.18	.21	.23	.26	.28	.30	.32	.34	.36	.38
19	.12	.15	.17	.20	.22	.25	.27	.29	.31	.33
20	.06	.09	.12	.14	.17	.19	.21	.24	.26	.28
21	.00	.03	.06	.09	.11	.14	.16	.18	.21	.23
22			.00	.03	.06	.08	.11	.13	.15	.18
23					.00	.03	.05	.08	.10	.13
24							.00	.03	.05	.08
25									.00	.03

TABLE I   Mean ranges of samples from a normal distribution

This table features the expected values (means) of the ranges of samples of varying sizes n from a standard normal distribution ($\mu = 0$, $\sigma = 1$, items normally distributed). The means are given, exact to three decimal places, for sample sizes $n = 2$ to $n = 200$ in increments of one, and for sample sizes $n = 200$ to $n = 1000$ in increments of 10. The function is in standard deviation units.

To find the mean range for a sample of size n, look up the function corresponding to the argument in the margin. For example, the mean range for a sample of $n = 173$ is 5.395, and for a sample of $n = 713$ it is 6.286 by linear interpolation between 6.283 and 6.292, the values for $n = 710$ and 720, respectively. Remember that the columns after $n = 200$ represent increments of 10.

The table of mean ranges is used in estimating population standard deviations from sample ranges (Section 4.9).

Values for this table have been rounded and taken from the more extensive table 27 (given to five decimal places) in E. S. Pearson and H. O. Hartley, *Biometrika Tables for Statisticians*, Vol. 1 (Cambridge University Press, 1958) with permission of the publishers.

TABLE **I**   Mean ranges of samples from a normal distribution

n	0	1	2	3	4	5	6	7	8	9
0			1.128	1.693	2.059	2.326	2.534	2.704	2.847	2.970
10	3.078	3.173	3.258	3.336	3.407	3.472	3.532	3.588	3.640	3.689
20	3.735	3.778	3.819	3.858	3.895	3.931	3.964	3.997	4.027	4.057
30	4.086	4.113	4.139	4.165	4.189	4.213	4.236	4.259	4.280	4.301
40	4.322	4.341	4.361	4.379	4.398	4.415	4.433	4.450	4.466	4.482
50	4.498	4.514	4.529	4.543	4.558	4.572	4.586	4.599	4.613	4.626
60	4.639	4.651	4.663	4.676	4.687	4.699	4.711	4.722	4.733	4.744
70	4.755	4.765	4.776	4.786	4.796	4.806	4.816	4.825	4.835	4.844
80	4.854	4.863	4.872	4.881	4.889	4.898	4.906	4.915	4.923	4.931
90	4.939	4.947	4.955	4.963	4.971	4.978	4.986	4.993	5.001	5.008
100	5.015	5.022	5.029	5.036	5.043	5.050	5.057	5.063	5.070	5.076
110	5.083	5.089	5.096	5.102	5.108	5.114	5.120	5.126	5.132	5.138
120	5.144	5.150	5.156	5.161	5.167	5.173	5.178	5.184	5.189	5.195
130	5.200	5.205	5.211	5.216	5.221	5.226	5.231	5.236	5.241	5.246
140	5.251	5.256	5.261	5.266	5.271	5.275	5.280	5.285	5.289	5.294
150	5.298	5.303	5.308	5.312	5.316	5.321	5.325	5.330	5.334	5.338
160	5.342	5.347	5.351	5.355	5.359	5.363	5.367	5.371	5.375	5.379
170	5.383	5.387	5.391	5.395	5.399	5.403	5.407	5.411	5.414	5.418
180	5.422	5.426	5.429	5.433	5.437	5.440	5.444	5.447	5.451	5.454
190	5.458	5.461	5.465	5.468	5.472	5.475	5.479	5.482	5.485	5.489

n	0	10	20	30	40	50	60	70	80	90
200	5.492	5.524	5.555	5.584	5.612	5.638	5.664	5.688	5.771	5.734
300	5.756	5.776	5.797	5.816	5.835	5.853	5.871	5.888	5.904	5.921
400	5.936	5.952	5.967	5.981	5.995	6.009	6.023	6.036	6.049	6.061
500	6.073	6.085	6.097	6.109	6.120	6.131	6.142	6.153	6.163	6.173
600	6.183	6.193	6.203	6.213	6.222	6.231	6.240	6.249	6.258	6.267
700	6.275	6.283	6.292	6.300	6.308	6.316	6.324	6.331	6.339	6.346
800	6.354	6.361	6.368	6.375	6.382	6.389	6.396	6.402	6.409	6.416
900	6.422	6.429	6.435	6.441	6.447	6.453	6.459	6.465	6.471	6.477
1000	6.483									

TABLE J Critical values of the studentized range

This table furnishes the critical values for the distribution of the ratio $Q_{\alpha[k,v]} = Range/s$, known as the studentized range. The range is computed over k variates, and the standard deviation s must be independently estimated and based upon v degrees of freedom. The critical values are tabulated for values of k, the number of items over which the range is computed, from 2 to 20 in increments of one, from 20 to 40 in increments of two, and from 40 to 100 in increments of 10. For degrees of freedom v, arguments are furnished from 1 to 20 in increments of one and for $v = 24, 30, 40, 60, 120$, and ∞. Note that these latter values of v lend themselves to harmonic interpolation. There are separate tables of the critical values of the studentized range representing the upper 5% and 1% points, respectively. The corresponding probabilities $\alpha = 0.05$ and 0.01 represent *one tail* of the distribution of Q.

To find the 1% critical values of the studentized range in a sample of 6 items but with a standard deviation independently obtained and based on 18 degrees of freedom (as in an analysis of variance with six samples of four items each), we enter the second part of the table ($\alpha = 0.01$) at $k = 6$ and $v = 18$, and find $Q_{.01[6,18]} = 5.603$.

The most frequent application of the studentized range is in unplanned multiple-comparisons tests in the analysis of variance (Section 9.7).

Values in this table have been copied from a more extensive one by H. L. Harter (*Ann. Math. Stat.* **31**:1122–1147, 1960) with permission of the publisher.

TABLE J Critical values of the studentized range $\alpha = 0.05$

ν \ k	2	3	4	5	6	7	8	9	10
1	17.97	26.98	32.82	37.08	40.41	43.12	45.40	47.36	49.07
2	6.085	8.331	9.798	10.88	11.75	12.44	13.03	13.54	13.99
3	4.501	5.910	6.825	7.502	8.037	8.478	8.853	9.177	9.462
4	3.927	5.040	5.757	6.287	6.707	7.053	7.347	7.602	7.826
5	3.635	4.602	5.218	5.673	6.033	6.330	6.582	6.802	6.995
6	3.461	4.339	4.896	5.305	5.628	5.895	6.122	6.319	6.493
7	3.344	4.165	4.681	5.060	5.359	5.606	5.815	5.998	6.158
8	3.261	4.041	4.529	4.886	5.167	5.399	5.597	5.767	5.918
9	3.199	3.949	4.415	4.756	5.024	5.244	5.432	5.595	5.739
10	3.151	3.877	4.327	4.654	4.912	5.124	5.305	5.461	5.599
11	3.113	3.820	4.256	4.574	4.823	5.028	5.202	5.353	5.487
12	3.082	3.773	4.199	4.508	4.751	4.950	5.119	5.265	5.395
13	3.055	3.735	4.151	4.453	4.690	4.885	5.049	5.192	5.318
14	3.033	3.702	4.111	4.407	4.639	4.829	4.990	5.131	5.254
15	3.014	3.674	4.076	4.367	4.595	4.782	4.940	5.077	5.198
16	2.998	3.649	4.046	4.333	4.557	4.741	4.897	5.031	5.150
17	2.984	3.628	4.020	4.303	4.524	4.705	4.858	4.991	5.108
18	2.971	3.609	3.997	4.277	4.495	4.673	4.824	4.956	5.071
19	2.960	3.593	3.977	4.253	4.469	4.645	4.794	4.924	5.038
20	2.950	3.578	3.958	4.232	4.445	4.620	4.768	4.896	5.008
24	2.919	3.532	3.901	4.166	4.373	4.541	4.684	4.807	4.915
30	2.888	3.486	3.845	4.102	4.302	4.464	4.602	4.720	4.824
40	2.858	3.442	3.791	4.039	4.232	4.389	4.521	4.635	4.735
60	2.829	3.399	3.737	3.977	4.163	4.314	4.441	4.550	4.646
120	2.800	3.356	3.685	3.917	4.096	4.241	4.363	4.468	4.560
∞	2.772	3.314	3.633	3.858	4.030	4.170	4.286	4.387	4.474

ν \ k	11	12	13	14	15	16	17	18	19
1	50.59	51.96	53.20	54.33	55.36	56.32	57.22	58.04	58.83
2	14.39	14.75	15.08	15.38	15.65	15.91	16.14	16.37	16.57
3	9.717	9.946	10.15	10.35	10.53	10.69	10.84	10.98	11.11
4	8.027	8.208	8.373	8.525	8.664	8.794	8.914	9.028	9.134
5	7.168	7.324	7.466	7.596	7.717	7.828	7.932	8.030	8.122
6	6.649	6.789	6.917	7.034	7.143	7.244	7.338	7.426	7.508
7	6.302	6.431	6.550	6.658	6.759	6.852	6.939	7.020	7.097
8	6.054	6.175	6.287	6.389	6.483	6.571	6.653	6.729	6.802
9	5.867	5.983	6.089	6.186	6.276	6.359	6.437	6.510	6.579
10	5.722	5.833	5.935	6.028	6.114	6.194	6.269	6.339	6.405
11	5.605	5.713	5.811	5.901	5.984	6.062	6.134	6.202	6.265
12	5.511	5.615	5.710	5.798	5.878	5.953	6.023	6.089	6.151
13	5.431	5.533	5.625	5.711	5.789	5.862	5.931	5.995	6.055
14	5.364	5.463	5.554	5.637	5.714	5.786	5.852	5.915	5.974
15	5.306	5.404	5.493	5.574	5.649	5.720	5.785	5.846	5.904
16	5.256	5.352	5.439	5.520	5.593	5.662	5.727	5.786	5.843
17	5.212	5.307	5.392	5.471	5.544	5.612	5.675	5.734	5.790
18	5.174	5.267	5.352	5.429	5.501	5.568	5.630	5.688	5.743
19	5.140	5.231	5.315	5.391	5.462	5.528	5.589	5.647	5.701
20	5.108	5.199	5.282	5.357	5.427	5.493	5.553	5.610	5.663
24	5.012	5.099	5.179	5.251	5.319	5.381	5.439	5.494	5.545
30	4.917	5.001	5.077	5.147	5.211	5.271	5.327	5.379	5.429
40	4.824	4.904	4.977	5.044	5.106	5.163	5.216	5.266	5.313
60	4.732	4.808	4.878	4.942	5.001	5.056	5.107	5.154	5.199
120	4.641	4.714	4.781	4.842	4.898	4.950	4.998	5.044	5.086
∞	4.552	4.622	4.685	4.743	4.796	4.845	4.891	4.934	4.974

TABLE J Critical values of the studentized range $\alpha = 0.05$

ν \ k	20	22	24	26	28	30	32	34	36
1	59.56	60.91	62.12	63.22	64.23	65.15	66.01	66.81	67.56
2	16.77	17.13	17.45	17.75	18.02	18.27	18.50	18.72	18.92
3	11.24	11.47	11.68	11.87	12.05	12.21	12.36	12.50	12.63
4	9.233	9.418	9.584	9.736	9.875	10.00	10.12	10.23	10.34
5	8.208	8.368	8.512	8.643	8.764	8.875	8.979	9.075	9.165
6	7.587	7.730	7.861	7.979	8.088	8.189	8.283	8.370	8.452
7	7.170	7.303	7.423	7.533	7.634	7.728	7.814	7.895	7.972
8	6.870	6.995	7.109	7.212	7.307	7.395	7.477	7.554	7.625
9	6.644	6.763	6.871	6.970	7.061	7.145	7.222	7.295	7.363
10	6.467	6.582	6.686	6.781	6.868	6.948	7.023	7.093	7.159
11	6.326	6.436	6.536	6.628	6.712	6.790	6.863	6.930	6.994
12	6.209	6.317	6.414	6.503	6.585	6.660	6.731	6.796	6.858
13	6.112	6.217	6.312	6.398	6.478	6.551	6.620	6.684	6.744
14	6.029	6.132	6.224	6.309	6.387	6.459	6.526	6.588	6.647
15	5.958	6.059	6.149	6.233	6.309	6.379	6.445	6.506	6.564
16	5.897	5.995	6.084	6.166	6.241	6.310	6.374	6.434	6.491
17	5.842	5.940	6.027	6.107	6.181	6.249	6.313	6.372	6.427
18	5.794	5.890	5.977	6.055	6.128	6.195	6.258	6.316	6.371
19	5.752	5.846	5.932	6.009	6.081	6.147	6.209	6.267	6.321
20	5.714	5.807	5.891	5.968	6.039	6.104	6.165	6.222	6.275
24	5.594	5.683	5.764	5.838	5.906	5.968	6.027	6.081	6.132
30	5.475	5.561	5.638	5.709	5.774	5.833	5.889	5.941	5.990
40	5.358	5.439	5.513	5.581	5.642	5.700	5.753	5.803	5.849
60	5.241	5.319	5.389	5.453	5.512	5.566	5.617	5.664	5.708
120	5.126	5.200	5.266	5.327	5.382	5.434	5.481	5.526	5.568
∞	5.012	5.081	5.144	5.201	5.253	5.301	5.346	5.388	5.427

ν \ k	38	40	50	60	70	80	90	100
1	68.26	68.92	71.73	73.97	75.82	77.40	78.77	79.98
2	19.11	19.28	20.05	20.66	21.16	21.59	21.96	22.29
3	12.75	12.87	13.36	13.76	14.08	14.36	14.61	14.82
4	10.44	10.53	10.93	11.24	11.51	11.73	11.92	12.09
5	9.250	9.330	9.674	9.949	10.18	10.38	10.54	10.69
6	8.529	8.601	8.913	9.163	9.370	9.548	9.702	9.839
7	8.043	8.110	8.400	8.632	8.824	8.989	9.133	9.261
8	7.693	7.756	8.029	8.248	8.430	8.586	8.722	8.843
9	7.428	7.488	7.749	7.958	8.132	8.281	8.410	8.526
10	7.220	7.279	7.529	7.730	7.897	8.041	8.166	8.276
11	7.053	7.110	7.352	7.546	7.708	7.847	7.968	8.075
12	6.916	6.970	7.205	7.394	7.552	7.687	7.804	7.909
13	6.800	6.854	7.083	7.267	7.421	7.552	7.667	7.769
14	6.702	6.754	6.979	7.159	7.309	7.438	7.550	7.650
15	6.618	6.669	6.888	7.065	7.212	7.339	7.449	7.546
16	6.544	6.594	6.810	6.984	7.128	7.252	7.360	7.457
17	6.479	6.529	6.741	6.912	7.054	7.176	7.283	7.377
18	6.422	6.471	6.680	6.848	6.989	7.109	7.213	7.307
19	6.371	6.419	6.626	6.792	6.930	7.048	7.152	7.244
20	6.325	6.373	6.576	6.740	6.877	6.994	7.097	7.187
24	6.181	6.226	6.421	6.579	6.710	6.822	6.920	7.008
30	6.037	6.080	6.267	6.417	6.543	6.650	6.744	6.827
40	5.893	5.934	6.112	6.255	6.375	6.477	6.566	6.645
60	5.750	5.789	5.958	6.093	6.206	6.303	6.387	6.462
120	5.607	5.644	5.802	5.929	6.035	6.126	6.205	6.275
∞	5.463	5.498	5.646	5.764	5.863	5.947	6.020	6.085

TABLE J Critical values of the studentized range $\alpha = 0.01$

ν\k	2	3	4	5	6	7	8	9	10
1	90.03	135.0	164.3	185.6	202.2	215.8	227.2	237.0	245.6
2	14.04	19.02	22.29	24.72	26.63	28.20	29.53	30.68	31.69
3	8.261	10.62	12.17	13.33	14.24	15.00	15.64	16.20	16.69
4	6.512	8.120	9.173	9.958	10.58	11.10	11.55	11.93	12.27
5	5.702	6.976	7.804	8.421	8.913	9.321	9.669	9.972	10.24
6	5.243	6.331	7.033	7.556	7.973	8.318	8.613	8.869	9.097
7	4.949	5.919	6.543	7.005	7.373	7.679	7.939	8.166	8.368
8	4.746	5.635	6.204	6.625	6.960	7.237	7.474	7.681	7.863
9	4.596	5.428	5.957	6.348	6.658	6.915	7.134	7.325	7.495
10	4.482	5.270	5.769	6.136	6.428	6.669	6.875	7.055	7.213
11	4.392	5.146	5.621	5.970	6.247	6.476	6.672	6.842	6.992
12	4.320	5.046	5.502	5.836	6.101	6.321	6.507	6.670	6.814
13	4.260	4.964	5.404	5.727	5.981	6.192	6.372	6.528	6.667
14	4.210	4.895	5.322	5.634	5.881	6.085	6.258	6.409	6.543
15	4.168	4.836	5.252	5.556	5.796	5.994	6.162	6.309	6.439
16	4.131	4.786	5.192	5.489	5.722	5.915	6.079	6.222	6.349
17	4.099	4.742	5.140	5.430	5.659	5.847	6.007	6.147	6.270
18	4.071	4.703	5.094	5.379	5.603	5.788	5.944	6.081	6.201
19	4.046	4.670	5.054	5.334	5.554	5.735	5.889	6.022	6.141
20	4.024	4.639	5.018	5.294	5.510	5.688	5.839	5.970	6.087
24	3.956	4.546	4.907	5.168	5.374	5.542	5.685	5.809	5.919
30	3.889	4.455	4.799	5.048	5.242	5.401	5.536	5.653	5.756
40	3.825	4.367	4.696	4.931	5.114	5.265	5.392	5.502	5.599
60	3.762	4.282	4.595	4.818	4.991	5.133	5.253	5.356	5.447
120	3.702	4.200	4.497	4.709	4.872	5.005	5.118	5.214	5.299
∞	3.643	4.120	4.403	4.603	4.757	4.882	4.987	5.078	5.157

ν\k	11	12	13	14	15	16	17	18	19
1	253.2	260.0	266.2	271.8	277.0	281.8	286.3	290.4	294.3
2	32.59	33.40	34.13	34.81	35.43	36.00	36.53	37.03	37.50
3	17.13	17.53	17.89	18.22	18.52	18.81	19.07	19.32	19.55
4	12.57	12.84	13.09	13.32	13.53	13.73	13.91	14.08	14.24
5	10.48	10.70	10.89	11.08	11.24	11.40	11.55	11.68	11.81
6	9.301	9.485	9.653	9.808	9.951	10.08	10.21	10.32	10.43
7	8.548	8.711	8.860	8.997	9.124	9.242	9.353	9.456	9.554
8	8.027	8.176	8.312	8.436	8.552	8.659	8.760	8.854	8.943
9	7.647	7.784	7.910	8.025	8.132	8.232	8.325	8.412	8.495
10	7.356	7.485	7.603	7.712	7.812	7.906	7.993	8.076	8.153
11	7.128	7.250	7.362	7.465	7.560	7.649	7.732	7.809	7.883
12	6.943	7.060	7.167	7.265	7.356	7.441	7.520	7.594	7.665
13	6.791	6.903	7.006	7.101	7.188	7.269	7.345	7.417	7.485
14	6.664	6.772	6.871	6.962	7.047	7.126	7.199	7.268	7.333
15	6.555	6.660	6.757	6.845	6.927	7.003	7.074	7.142	7.204
16	6.462	6.564	6.658	6.744	6.823	6.898	6.967	7.032	7.093
17	6.381	6.480	6.572	6.656	6.734	6.806	6.873	6.937	6.997
18	6.310	6.407	6.497	6.579	6.655	6.725	6.792	6.854	6.912
19	6.247	6.342	6.430	6.510	6.585	6.654	6.719	6.780	6.837
20	6.191	6.285	6.371	6.450	6.523	6.591	6.654	6.714	6.771
24	6.017	6.106	6.186	6.261	6.330	6.394	6.453	6.510	6.563
30	5.849	5.932	6.008	6.078	6.143	6.203	6.259	6.311	6.361
40	5.686	5.764	5.835	5.900	5.961	6.017	6.069	6.119	6.165
60	5.528	5.601	5.667	5.728	5.785	5.837	5.886	5.931	5.974
120	5.375	5.443	5.505	5.562	5.614	5.662	5.708	5.750	5.790
∞	5.227	5.290	5.348	5.400	5.448	5.493	5.535	5.574	5.611

TABLE J Critical values of the studentized range $\alpha = 0.01$

$\nu \backslash k$	20	22	24	26	28	30	32	34	36
1	298.0	304.7	310.8	316.3	321.3	326.0	330.3	334.3	338.0
2	37.95	38.76	39.49	40.15	40.76	41.32	41.84	42.33	42.78
3	19.77	20.17	20.53	20.86	21.16	21.44	21.70	21.95	22.17
4	14.40	14.68	14.93	15.16	15.37	15.57	15.75	15.92	16.08
5	11.93	12.16	12.36	12.54	12.71	12.87	13.02	13.15	13.28
6	10.54	10.73	10.91	11.06	11.21	11.34	11.47	11.58	11.69
7	9.646	9.815	9.970	10.11	10.24	10.36	10.47	10.58	10.67
8	9.027	9.182	9.322	9.450	9.569	9.678	9.779	9.874	9.964
9	8.573	8.717	8.847	8.966	9.075	9.177	9.271	9.360	9.443
10	8.226	8.361	8.483	8.595	8.698	8.794	8.883	8.966	9.044
11	7.952	8.080	8.196	8.303	8.400	8.491	8.575	8.654	8.728
12	7.731	7.853	7.964	8.066	8.159	8.246	8.327	8.402	8.473
13	7.548	7.665	7.772	7.870	7.960	8.043	8.121	8.193	8.262
14	7.395	7.508	7.611	7.705	7.792	7.873	7.948	8.018	8.084
15	7.264	7.374	7.474	7.566	7.650	7.728	7.800	7.869	7.932
16	7.152	7.258	7.356	7.445	7.527	7.602	7.673	7.739	7.802
17	7.053	7.158	7.253	7.340	7.420	7.493	7.563	7.627	7.687
18	6.968	7.070	7.163	7.247	7.325	7.398	7.465	7.528	7.587
19	6.891	6.992	7.082	7.166	7.242	7.313	7.379	7.440	7.498
20	6.823	6.922	7.011	7.092	7.168	7.237	7.302	7.362	7.419
24	6.612	6.705	6.789	6.865	6.936	7.001	7.062	7.119	7.173
30	6.407	6.494	6.572	6.644	6.710	6.772	6.828	6.881	6.932
40	6.209	6.289	6.362	6.429	6.490	6.547	6.600	6.650	6.697
60	6.015	6.090	6.158	6.220	6.277	6.330	6.378	6.424	6.467
120	5.827	5.897	5.959	6.016	6.069	6.117	6.162	6.204	6.244
∞	5.645	5.709	5.766	5.818	5.866	5.911	5.952	5.990	6.026

$\nu \backslash k$	38	40	50	60	70	80	90	100
1	341.5	344.8	358.9	370.1	379.4	387.3	394.1	400.1
2	43.21	43.61	45.33	46.70	47.83	48.80	49.64	50.38
3	22.39	22.59	23.45	24.13	24.71	25.19	25.62	25.99
4	16.23	16.37	16.98	17.46	17.86	18.20	18.50	18.77
5	13.40	13.52	14.00	14.39	14.72	14.99	15.23	15.45
6	11.80	11.90	12.31	12.65	12.92	13.16	13.37	13.55
7	10.77	10.85	11.23	11.52	11.77	11.99	12.17	12.34
8	10.05	10.13	10.47	10.75	10.97	11.17	11.34	11.49
9	9.521	9.594	9.912	10.17	10.38	10.57	10.73	10.87
10	9.117	9.187	9.486	9.726	9.927	10.10	10.25	10.39
11	8.798	8.864	9.148	9.377	9.568	9.732	9.875	10.00
12	8.539	8.603	8.875	9.094	9.277	9.434	9.571	9.693
13	8.326	8.387	8.648	8.859	9.035	9.187	9.318	9.436
14	8.146	8.204	8.457	8.661	8.832	8.978	9.106	9.219
15	7.992	8.049	8.295	8.492	8.658	8.800	8.924	9.035
16	7.860	7.916	8.154	8.347	8.507	8.646	8.767	8.874
17	7.745	7.799	8.031	8.219	8.377	8.511	8.630	8.735
18	7.643	7.696	7.924	8.107	8.261	8.393	8.508	8.611
19	7.553	7.605	7.828	8.008	8.159	8.288	8.401	8.502
20	7.473	7.523	7.742	7.919	8.067	8.194	8.305	8.404
24	7.223	7.270	7.476	7.642	7.780	7.900	8.004	8.097
30	6.978	7.023	7.215	7.370	7.500	7.611	7.709	7.796
40	6.740	6.782	6.960	7.104	7.225	7.328	7.419	7.500
60	6.507	6.546	6.710	6.843	6.954	7.050	7.133	7.207
120	6.281	6.316	6.467	6.588	6.689	6.776	6.852	6.919
∞	6.060	6.092	6.228	6.338	6.429	6.507	6.575	6.636

TABLE **K**   Critical values for Welsch's step-up procedure

This table furnishes critical values for Welsch's step-up procedure (described by him as the GAPA procedure). The table is entered via two arguments. The first of these is k, the number of means in the set being compared, and the second is j, the "stretch" or size of the subset j in k being tested for homogeneity. Both j and k are furnished from 2 to 10 in increments of one. The values in the table are actually studentized ranges, $Q_{\alpha j}$, for unusual probabilities $\alpha_j = \alpha(j/k)$, where α_j is the type I error for testing a range of j adjacent means. The values are given for 5% experimentwise error rates and are tabled for degrees of freedom v from 5 to 16 in increments of one, and for $v = 18, 20, 24, 30, 40, 60, 120$, and ∞. For degrees of freedom other than those tabled, employ harmonic interpolation.

For $k = 6$ and $j = 3$, the critical value at 40 degrees of freedom according to the table is 3.89. This critical value is employed in Welsch's step-up procedure for multiple comparisons of means (Section 9.7).

This table is copied in rearranged form from a technical report by R. E. Welsch, *Tables for Stepwise Multiple Comparison Procedures* (Working Paper no. 949–77. Sloan School of Management, 1977) with permission of the author.

TABLE **K**   Critical values for Welsch's step-up procedure

$\nu = 5$

$j \backslash k$	2	3	4	5	6	7	8	9	10
2	3.64	3.64	4.47	4.76	5.00	5.21	5.39	5.55	5.70
3		5.38	4.60	5.37	5.64	5.86	6.06	6.24	6.41
4			5.46	5.37	5.84	6.07	6.28	6.47	6.64
5				5.72	5.84	6.21	6.42	6.61	6.79
6					6.05	6.21	6.50	6.70	6.88
7						6.34	6.50	6.75	6.94
8							6.59	6.75	6.97
9								6.80	6.97
10									7.00

$\nu = 6$

$j \backslash k$	2	3	4	5	6	7	8	9	10
2	3.46	3.46	4.20	4.44	4.65	4.83	4.98	5.12	5.24
3		4.95	4.34	5.01	5.23	5.42	5.58	5.73	5.87
4			5.10	5.01	5.42	5.62	5.80	5.95	6.09
5				5.36	5.42	5.75	5.93	6.09	6.24
6					5.65	5.75	6.02	6.19	6.33
7						5.91	6.02	6.25	6.40
8							6.13	6.25	6.44
9								6.32	6.44
10									6.50

$\nu = 7$

$j \backslash k$	2	3	4	5	6	7	8	9	10
2	3.34	3.34	4.02	4.24	4.42	4.58	4.72	4.84	4.95
3		4.69	4.17	4.76	4.96	5.13	5.27	5.40	5.52
4			4.86	4.76	5.15	5.33	5.48	5.61	5.74
5				5.11	5.15	5.46	5.61	5.75	5.88
6					5.39	5.46	5.71	5.85	5.98
7						5.62	5.71	5.92	6.04
8							5.82	5.92	6.10
9								6.00	6.10
10									6.17

$\nu = 8$

$j \backslash k$	2	3	4	5	6	7	8	9	10
2	3.26	3.26	3.89	4.10	4.27	4.41	4.53	4.65	4.75
3		4.51	4.05	4.59	4.77	4.92	5.06	5.17	5.28
4			4.69	4.59	4.96	5.12	5.26	5.38	5.49
5				4.94	4.96	5.25	5.39	5.51	5.63
6					5.19	5.25	5.48	5.61	5.72
7						5.41	5.48	5.68	5.79
8							5.60	5.68	5.85
9								5.77	5.85
10									5.92

TABLE **K**   Critical values for Welsch's step-up procedure (*continued*)

$\nu = 9$

$j \backslash k$	2	3	4	5	6	7	8	9	10
2	3.20	3.20	3.80	3.99	4.15	4.28	4.40	4.50	4.60
3		4.37	3.95	4.47	4.63	4.77	4.90	5.01	5.10
4			4.56	4.47	4.82	4.96	5.09	5.20	5.30
5				4.81	4.82	5.09	5.22	5.34	5.44
6					5.05	5.09	5.31	5.43	5.54
7						5.26	5.31	5.50	5.61
8							5.44	5.50	5.67
9								5.60	5.67
10									5.74

$\nu = 10$

$j \backslash k$	2	3	4	5	6	7	8	9	10
2	3.15	3.15	3.73	3.91	4.06	4.19	4.30	4.39	4.48
3		4.27	3.88	4.37	4.53	4.66	4.77	4.88	4.97
4			4.47	4.37	4.71	4.84	4.96	5.07	5.16
5				4.71	4.71	4.97	5.09	5.20	5.30
6					4.94	4.97	5.18	5.30	5.39
7						5.14	5.18	5.37	5.47
8							5.32	5.37	5.52
9								5.47	5.52
10									5.60

$\nu = 11$

$j \backslash k$	2	3	4	5	6	7	8	9	10
2	3.11	3.11	3.67	3.84	3.99	4.11	4.22	4.31	4.39
3		4.19	3.83	4.29	4.44	4.57	4.68	4.77	4.86
4			4.39	4.29	4.62	4.75	4.86	4.96	5.05
5				4.63	4.62	4.87	4.99	5.09	5.18
6					4.85	4.87	5.08	5.19	5.28
7						5.04	5.08	5.26	5.35
8							5.21	5.26	5.41
9								5.36	5.41
10									5.49

$\nu = 12$

$j \backslash k$	2	3	4	5	6	7	8	9	10
2	3.08	3.08	3.62	3.79	3.93	4.05	4.15	4.24	4.32
3		4.13	3.78	4.23	4.37	4.49	4.60	4.69	4.77
4			4.32	4.23	4.55	4.67	4.78	4.88	4.96
5				4.56	4.55	4.80	4.91	5.00	5.09
6					4.78	4.80	5.00	5.10	5.18
7						4.97	5.00	5.17	5.26
8							5.13	5.17	5.31
9								5.27	5.31
10									5.40

TABLE K Critical values for Welsch's step-up procedure (*continued*)

$\nu = 13$

$j \backslash k$	2	3	4	5	6	7	8	9	10
2	3.06	3.06	3.58	3.75	3.88	4.00	4.10	4.18	4.26
3		4.07	3.74	4.18	4.32	4.43	4.53	4.62	4.70
4			4.27	4.18	4.49	4.61	4.71	4.80	4.89
5				4.51	4.49	4.73	4.84	4.93	5.01
6					4.72	4.73	4.93	5.02	5.11
7						4.90	4.93	5.09	5.18
8							5.06	5.09	5.24
9								5.20	5.24
10									5.33

$\nu = 14$

$j \backslash k$	2	3	4	5	6	7	8	9	10
2	3.03	3.03	3.55	3.71	3.84	3.95	4.05	4.14	4.21
3		4.03	3.71	4.13	4.27	4.38	4.48	4.56	4.64
4			4.23	4.13	4.44	4.56	4.66	4.74	4.82
5				4.46	4.44	4.68	4.78	4.87	4.95
6					4.67	4.68	4.87	4.96	5.04
7						4.85	4.87	5.03	5.11
8							5.00	5.03	5.17
9								5.14	5.17
10									5.26

$\nu = 15$

$j \backslash k$	2	3	4	5	6	7	8	9	10
2	3.01	3.01	3.52	3.68	3.81	3.92	4.01	4.09	4.17
3		3.99	3.68	4.10	4.23	4.34	4.43	4.52	4.59
4			4.19	4.10	4.40	4.51	4.61	4.69	4.77
5				4.42	4.40	4.63	4.73	4.82	4.89
6					4.62	4.63	4.82	4.91	4.99
7						4.80	4.82	4.98	5.06
8							4.95	4.98	5.12
9								5.09	5.12
10									5.20

$\nu = 16$

$j \backslash k$	2	3	4	5	6	7	8	9	10
2	3.00	3.00	3.50	3.65	3.78	3.89	3.98	4.06	4.13
3		3.96	3.66	4.06	4.19	4.30	4.39	4.47	4.55
4			4.16	4.06	4.36	4.47	4.57	4.65	4.72
5				4.39	4.36	4.59	4.69	4.77	4.85
6					4.59	4.59	4.78	4.86	4.94
7						4.76	4.78	4.93	5.01
8							4.91	4.93	5.07
9								5.04	5.07
10									5.16

TABLE K Critical values for Welsch's step-up procedure (*continued*)

$\nu = 18$

$j \backslash k$	2	3	4	5	6	7	8	9	10
2	2.97	2.97	3.46	3.61	3.73	3.84	3.92	4.00	4.07
3		3.91	3.62	4.01	4.13	4.24	4.33	4.40	4.47
4			4.10	4.01	4.30	4.40	4.50	4.58	4.65
5				4.33	4.30	4.52	4.62	4.70	4.77
6					4.52	4.52	4.70	4.79	4.86
7						4.69	4.70	4.86	4.93
8							4.83	4.86	4.99
9								4.96	4.99
10									5.08

$\nu = 20$

$j \backslash k$	2	3	4	5	6	7	8	9	10
2	2.95	2.95	3.43	3.58	3.70	3.80	3.88	3.96	4.02
3		3.87	3.58	3.97	4.09	4.19	4.28	4.35	4.42
4			4.06	3.97	4.25	4.35	4.44	4.52	4.59
5				4.29	4.25	4.47	4.56	4.64	4.71
6					4.47	4.47	6.65	4.73	4.80
7						4.64	4.65	4.80	4.87
8							4.78	4.80	4.92
9								4.90	4.92
10									5.01

$\nu = 24$

$j \backslash k$	2	3	4	5	6	7	8	9	10
2	2.92	2.92	3.38	3.52	3.64	3.74	3.82	3.89	3.96
3		3.80	3.54	3.91	4.02	4.12	4.20	4.27	4.34
4			4.00	3.91	4.18	4.28	4.36	4.44	4.50
5				4.22	4.18	4.39	4.48	4.55	4.62
6					4.40	4.39	4.56	4.64	4.71
7						4.56	4.56	4.71	4.78
8							4.69	4.71	4.83
9								4.81	4.83
10									4.92

$\nu = 30$

$j \backslash k$	2	3	4	5	6	7	8	9	10
2	2.89	2.89	3.34	3.48	3.59	3.68	3.76	3.83	3.89
3		3.74	3.49	3.85	3.96	4.05	4.13	4.20	4.26
4			3.94	3.85	4.11	4.21	4.29	4.36	4.42
5				4.15	4.11	4.32	4.40	4.47	4.53
6					4.33	4.32	4.48	4.55	4.62
7						4.48	4.48	4.62	4.68
8							4.61	4.62	4.74
9								4.73	4.74
10									4.83

TABLE K Critical values for Welsch's step-up procedure (*continued*)

$\nu = 40$

$j \backslash k$	2	3	4	5	6	7	8	9	10
2	2.86	2.86	3.29	3.43	3.53	3.62	3.70	3.77	3.82
3		3.69	3.45	3.79	3.89	3.98	4.06	4.12	4.18
4			3.88	3.79	4.05	4.14	4.21	4.28	4.34
5				4.09	4.05	4.24	4.32	4.39	4.45
6					4.26	4.24	4.40	4.47	4.53
7						4.40	4.40	4.54	4.60
8							4.53	4.54	4.65
9								4.64	4.65
10									4.74

$\nu = 60$

$j \backslash k$	2	3	4	5	6	7	8	9	10
2	2.83	2.83	3.25	3.38	3.48	3.57	3.64	3.71	3.76
3		3.63	3.41	3.73	3.83	3.91	3.99	4.05	4.10
4			3.82	3.74	3.98	4.07	4.14	4.20	4.26
5				4.02	3.98	4.17	4.24	4.31	4.36
6					4.19	4.17	4.33	4.39	4.45
7						4.33	4.33	4.45	4.51
8							4.45	4.45	4.56
9								4.56	4.56
10									4.65

$\nu = 120$

$j \backslash k$	2	3	4	5	6	7	8	9	10
2	2.80	2.80	3.21	3.33	3.43	3.52	3.59	3.65	3.70
3		3.58	3.36	3.68	3.77	3.85	3.92	3.98	4.03
4			3.76	3.69	3.92	4.00	4.07	4.13	4.18
5				3.96	3.92	4.10	4.17	4.23	4.28
6					4.12	4.10	4.25	4.31	4.36
7						4.26	4.25	4.37	4.43
8							4.37	4.37	4.48
9								4.48	4.48
10									4.57

$\nu = \infty$

$j \backslash k$	2	3	4	5	6	7	8	9	10
2	2.77	2.77	3.17	3.29	3.39	3.46	3.53	3.59	3.64
3		3.52	3.32	3.62	3.71	3.79	3.85	3.91	3.96
4			3.71	3.64	3.86	3.93	4.00	4.05	4.10
5				3.90	3.86	4.03	4.10	4.15	4.20
6					4.06	4.03	4.18	4.23	4.28
7						4.19	4.18	4.29	4.34
8							4.30	4.29	4.39
9								4.39	4.39
10									4.48

TABLE **L**   Critical values of the studentized augmented range

This table furnishes critical values for the studentized augmented range $Q'_{\alpha[k,v]}$ employed in the Spjøtvoll–Stoline T'-method of multiple comparisons. Values are tabled for $k = 2$ to 8 means in increments of one and for $v = 5, 7, 10, 12, 16, 20, 24, 30, 40, 60, 120,$ and ∞ degrees of freedom. Three experimentwise error rates, α—0.10, 0.05, and 0.01—are given for each combination of k and v. Harmonic interpolation is recommended for degrees of freedom that are not furnished.

To find the 5% critical value of the studentized augmented range over $k = 5$ means at 20 degrees of freedom, enter the table at $k = 5$, $v = 20$, and $\alpha = .05$ to obtain $Q'_{.05[5,20]} = 4.233$. For $k > 8$, the studentized range furnished in Table **J** may be used because for $k > 8$ these two distributions are very similar.

This table is used for the Spjøtvoll–Stoline T'-test (Section 9.7). It can also be employed in multiple comparisons of regression coefficients (Box 14.8, Section 14.8).

This table has been copied and rearranged from the tables furnished by M. R. Stoline (*J. Amer. Stat. Assn.* **73:**656–660, 1978) with permission of the publisher.

TABLE **L**   Critical values of the studentized augmented range

$$k$$

ν	α	2	3	4	5	6	7	8
5	.10	3.060	3.772	4.282	4.671	4.982	5.239	5.458
	.05	3.832	4.654	5.236	5.680	6.036	6.331	6.583
	.01	5.903	7.030	7.823	8.429	8.916	9.322	9.669
7	.10	2.848	3.491	3.943	4.285	4.556	4.781	4.972
	.05	3.486	4.198	4.692	5.064	5.360	5.606	5.816
	.01	5.063	5.947	6.551	7.008	7.374	7.679	7.939
10	.10	2.704	3.300	3.712	4.021	4.265	4.466	4.636
	.05	3.259	3.899	4.333	4.656	4.913	5.124	5.305
	.01	4.550	5.284	5.773	6.138	6.428	6.669	6.875
12	.10	2.651	3.230	3.628	3.924	4.157	4.349	4.511
	.05	3.177	3.791	4.204	4.509	4.751	4.950	5.119
	.01	4.373	5.056	5.505	5.837	6.101	6.321	6.507
16	.10	2.588	3.146	3.526	3.806	4.027	4.207	4.360
	.05	3.080	3.663	4.050	4.334	4.557	4.741	4.897
	.01	4.169	4.792	5.194	5.489	5.722	5.915	6.079
20	.10	2.551	3.097	3.466	3.738	3.950	4.124	4.271
	.05	3.024	3.590	3.961	4.233	4.446	4.620	4.768
	.01	4.055	4.644	5.019	5.294	5.510	5.688	5.839
24	.10	2.527	3.065	3.427	3.693	3.901	4.070	4.213
	.05	2.988	3.542	3.904	4.167	4.373	4.541	4.684
	.01	3.982	4.549	4.908	5.169	5.374	5.542	5.685
30	.10	2.503	3.034	3.389	3.649	3.851	4.016	4.155
	.05	2.952	3.496	3.847	4.103	4.302	4.464	4.602
	.01	3.912	4.458	4.800	5.048	5.242	5.401	5.536
40	.10	2.480	3.003	3.352	3.605	3.803	3.963	4.099
	.05	2.918	3.450	3.792	4.040	4.232	4.389	4.521
	.01	3.844	4.370	4.696	4.931	5.115	5.265	5.392
60	.10	2.457	2.972	3.315	3.563	3.755	3.911	4.042
	.05	2.884	3.406	3.738	3.978	4.163	4.314	4.441
	.01	3.778	4.284	4.595	4.818	4.991	5.133	5.253
120	.10	2.434	2.943	3.278	3.520	3.707	3.859	3.987
	.05	2.851	3.362	3.686	3.917	4.096	4.241	4.363
	.01	3.714	4.201	4.497	4.709	4.872'	5.005	5.118
∞	.10	2.412	2.913	3.243	3.479	3.661	3.808	3.931
	.05	2.819	3.320	3.634	3.858	4.030	4.170	4.286
	.01	3.653	4.121	4.403	4.603	4.757	4.882	4.987

TABLE **M** Critical values of the studentized maximum modulus distribution

This table furnishes values of the studentized maximum modulus $m_{\alpha[k^*,v]}$, which is used in Hochberg's GT2-method for multiple comparisons of means. For arguments k (the number of means) ranging from $k = 3$ to $k = 20$ in increments of one, critical values are given for degrees of freedom $v = 5, 7, 10, 12, 16, 20,$ $24, 30, 40, 60, 120,$ and ∞. For k ranging from 20 to 50 in increments of two, from 50 to 80 in increments of 10, and for $k = 100$, critical values are given for degrees of freedom v ranging from 20 to 40 in increments of two; from 40 to 60 in increments of 10; from 60 to 120 in increments of 20; and for 240, 480, and ∞. Three experimentwise type I error rates are furnished: $\alpha = 0.10, 0.05,$ and 0.01.

For a set of k means, the GT2-method permits up to $k^* = k(k - 1)/2$ pairwise comparisons. Although the proper parameter of m is k^*, the number of comparisons intended, we precede it by the more convenient k. Interpolation for degrees of freedom v in this table is harmonic. If only a subset of k^* (all possible pairwise tests) is intended, one can interpolate for the studentized maximum modulus corresponding to the intended number of comparisons k' as follows. Interpolate linearly for $m_{\alpha[k^*,v]}$ between the bracketing values of k^* (*not* the corresponding arguments k) and between their reciprocals. (The second interpolation corresponds to a harmonic interpolation.) The desired interpolated value of m is the average of the linear and the harmonic interpolated value. A conservative rule is to use the value of $m_{\alpha[k^*,v]}$ corresponding to the upper bracketing value of k^*.

Looking up the 5% critical value of m for 24 degrees of freedom and a set of $k = 10$ means, which corresponds to $k^* = 45$, we obtain $m_{.05[45,24]} = 3.641$. These critical values are for two-tailed tests.

This table is used in the GT2-method of multiple comparisons of means (Section 9.7). It can also be employed in multiple comparisons of regression coefficients (Box 14.8, Section 14.8).

This table has been copied in rearranged form from tables by M. R. Stoline and H. K. Ury (*Technometrics* **21**:87–93, 1979) and by H. K. Ury, M. R. Stoline, and B. T. Mitchell [*Commun. Stat.* **B9**(2), 1980] with permission of the publishers.

TABLE M Critical values of the studentized maximum modulus distribution

Degrees of freedom v

k	k^*	α	5	7	10	12	16	20
3	3	.10	2.769	2.555	2.410	2.357	2.293	2.255
		.05	3.399	3.055	2.829	2.747	2.650	2.594
		.01	5.106	4.296	3.801	3.631	3.434	3.323
4	6	.10	3.239	2.961	2.771	2.701	2.616	2.567
		.05	3.928	3.489	3.199	3.095	2.969	2.897
		.01	5.812	4.814	4.205	3.995	3.753	3.617
5	10	.10	3.576	3.253	3.029	2.946	2.845	2.786
		.05	4.312	3.805	3.467	3.345	3.199	3.114
		.01	6.334	5.198	4.503	4.263	3.986	3.831
6	15	.10	3.837	3.478	3.229	3.136	3.022	2.956
		.05	4.610	4.051	3.677	3.540	3.377	3.282
		.01	6.744	5.502	4.739	4.475	4.170	3.999
7	21	.10	4.048	3.661	3.391	3.290	3.166	3.093
		.05	4.853	4.252	3.848	3.700	3.522	3.419
		.01	7.079	5.752	4.933	4.650	4.322	4.137
8	28	.10	4.224	3.814	3.527	3.419	3.286	3.208
		.05	5.057	4.421	3.992	3.835	3.645	3.534
		.01	7.362	5.963	5.099	4.799	4.451	4.255
9	36	.10	4.375	3.945	3.644	3.530	3.390	3.306
		.05	5.232	4.566	4.116	3.951	3.751	3.634
		.01	7.605	6.146	5.242	4.929	4.563	4.357
10	45	.10	4.506	4.060	3.746	3.627	3.480	3.393
		.05	5.384	4.693	4.225	4.053	3.844	3.721
		.01	7.819	6.307	5.369	5.043	4.662	4.447
11	55	.10	4.623	4.162	3.836	3.713	3.560	3.469
		.05	5.520	4.806	4.322	4.143	3.926	3.799
		.01	8.008	6.450	5.482	5.145	4.750	4.528
12	66	.10	4.727	4.252	3.917	3.790	3.633	3.538
		.05	5.641	4.907	4.409	4.225	4.001	3.869
		.01	8.178	6.579	5.584	5.237	4.830	4.601
13	78	.10	4.821	4.335	3.991	3.860	3.698	3.601
		.05	5.750	4.999	4.488	4.299	4.068	3.933
		.01	8.332	6.696	5.677	5.321	4.903	4.667
14	91	.10	4.906	4.410	4.058	3.924	3.758	3.658
		.05	5.850	5.083	4.560	4.367	4.130	3.991
		.01	8.473	6.803	5.762	5.398	4.970	4.728
15	105	.10	4.985	4.479	4.119	3.983	3.813	3.710
		.05	5.942	5.160	4.627	4.429	4.187	4.045
		.01	8.602	6.902	5.841	5.469	5.032	4.785
16	120	.10	5.058	4.542	4.176	4.037	3.863	3.759
		.05	6.027	5.232	4.688	4.487	4.240	4.095
		.01	8.722	6.994	5.914	5.535	5.090	4.837
17	136	.10	5.125	4.601	4.229	4.087	3.911	3.804
		.05	6.106	5.298	4.746	4.540	4.289	4.141
		.01	8.833	7.079	5.982	5.597	5.144	4.886
18	153	.10	5.188	4.657	4.279	4.135	3.955	3.846
		.05	6.179	5.360	4.799	4.591	4.335	4.184
		.01	8.937	7.158	6.045	5.654	5.194	4.932
19	171	.10	5.247	4.708	4.325	4.179	3.996	3.886
		.05	6.248	5.418	4.849	4.638	4.379	4.225
		.01	9.034	7.233	6.105	5.708	5.242	4.975
20	190	.10	5.302	4.757	4.369	4.220	4.035	3.923
		.05	6.313	5.472	4.897	4.682	4.419	4.264
		.01	9.126	7.303	6.161	5.760	5.286	5.016

Degrees of freedom v

24	30	40	60	120	∞	k	k^*	α
2.231	2.207	2.183	2.160	2.137	2.114	3	3	.10
2.558	2.522	2.488	2.454	2.420	2.388			.05
3.253	3.185	3.119	3.055	2.993	2.934			.01
2.534	2.502	2.470	2.439	2.408	2.378	4	6	.10
2.851	2.805	2.760	2.716	2.673	2.631			.05
3.531	3.447	3.367	3.289	3.215	3.143			.01
2.747	2.709	2.671	2.633	2.596	2.560	5	10	.10
3.059	3.005	2.952	2.900	2.849	2.800			05
3.732	3.637	3.545	3.456	3.371	3.289			.01
2.911	2.868	2.825	2.782	2.739	2.697	6	15	.10
3.220	3.160	3.100	3.041	2.984	2.928			.05
3.890	3.785	3.683	3.586	3.492	3.402			.01
3.044	2.996	2.948	2.901	2.854	2.807	7	21	.10
3.352	3.285	3.220	3.156	3.093	3.031			.05
4.020	3.906	3.796	3.691	3.590	3.493			.01
3.156	3.104	3.052	3.001	2.949	2.898	8	28	.10
3.462	3.391	3.321	3.251	3.183	3.117			.05
4.130	4.009	3.892	3.780	3.672	3.569			.01
3.251	3.196	3.141	3.086	3.031	2.976	9	36	.10
3.557	3.482	3.407	3.334	3.261	3.190			.05
4.225	4.098	3.975	3.857	3.743	3.634			.01
3.335	3.277	3.219	3.160	3.102	3.043	10	45	.10
3.641	3.562	3.483	3.406	3.329	3.254			.05
4.309	4.176	4.048	3.924	3.805	3.691			01
3.409	3.348	3.287	3.226	3.165	3.103	11	55	.10
3.716	3.633	3.551	3.470	3.389	3.310			.05
4.385	4.247	4.113	3.984	3.860	3.742			.01
3.476	3.413	3.349	3.286	3.221	3.157	12	66	.10
3.783	3.697	3.611	3.527	3.443	3.361			.05
4.453	4.310	4.172	4.038	3.910	3.787			.01
3.536	3.471	3.405	3.339	3.272	3.205	13	78	.10
3.843	3.755	3.667	3.579	3.492	3.407			.05
4.515	4.368	4.225	4.088	3.955	3.829			.01
3.591	3.524	3.456	3.388	3.319	3.249	14	91	.10
3.899	3.808	3.717	3.627	3.537	3.449			.05
4.572	4.421	4.275	4.133	3.997	3.867			.01
3.642	3.573	3.504	3.433	3.362	3.290	15	105	.10
3.951	3.857	3.764	3.671	3.578	3.487			.05
4.625	4.470	4.320	4.175	4.035	3.901			.01
3.689	3.619	3.547	3.475	3.402	3.327	16	120	.10
3.998	3.902	3.807	3.712	3.617	3.523			.05
4.674	4.516	4.362	4.214	4.070	3.934			.01
3.733	3.661	3.588	3.514	3.438	3.362	17	136	.10
4.043	3.945	3.847	3.750	3.652	3.556			.05
4.720	4.559	4.402	4.250	4.103	3.963			.01
3.774	3.700	3.626	3.550	3.473	3.394	18	153	.10
4.084	3.984	3.885	3.785	3.685	3.587			.05
4.763	4.599	4.439	4.284	4.134	3.991			.01
3.812	3.737	3.661	3.584	3.505	3.425	19	171	.10
4.123	4.022	3.920	3.818	3.717	3.615			.05
4.804	4.636	4.473	4.316	4.163	4.018			.01
3.848	3.772	3.695	3.616	3.536	3.453	20	190	.10
4.160	4.057	3.953	3.850	3.746	3.643			.05
4.842	4.672	4.506	4.346	4.191	4.043			.01

TABLE **M** Critical values of the studentized maximum modulus distribution (*continued*)

Degrees of freedom *v*

k	k*	α	20	22	24	26	28	30	32	34	36	38
20	190	.10	3.923	3.882	3.848	3.819	3.794	3.772	3.753	3.736	3.721	3.707
		.05	4.264	4.207	4.160	4.120	4.086	4.057	4.031	4.008	3.988	3.970
		.01	5.016	4.921	4.842	4.776	4.720	4.672	4.630	4.593	4.561	4.532
22	231	.10	3.992	3.950	3.914	3.884	3.859	3.836	3.816	3.799	3.783	3.769
		.05	4.334	4.276	4.228	4.187	4.152	4.121	4.095	4.071	4.050	4.031
		.01	5.092	4.993	4.912	4.845	4.787	4.737	4.694	4.657	4.623	4.594
24	276	.10	4.054	4.010	3.974	3.943	3.917	3.894	3.874	3.855	3.839	3.825
		.05	4.398	4.339	4.289	4.247	4.211	4.180	4.152	4.128	4.107	4.087
		.01	5.160	5.059	4.976	4.907	4.848	4.797	4.753	4.714	4.680	4.649
26	325	.10	4.110	4.066	4.029	3.997	3.970	3.946	3.926	3.907	3.891	3.876
		.05	4.456	4.395	4.345	4.302	4.265	4.233	4.205	4.180	4.158	4.138
		.01	5.222	5.119	5.035	4.964	4.903	4.851	4.806	4.766	4.731	4.700
28	378	.10	4.161	4.116	4.079	4.046	4.019	3.995	3.973	3.954	3.938	3.922
		.05	4.509	4.447	4.396	4.352	4.314	4.282	4.253	4.228	4.205	4.185
		.01	5.280	5.175	5.088	5.016	4.954	4.901	4.855	4.815	4.779	4.747
30	435	.10	4.209	4.163	4.124	4.092	4.064	4.039	4.017	3.998	3.981	3.965
		.05	4.559	4.496	4.443	4.398	4.360	4.327	4.298	4.272	4.249	4.228
		.01	5.333	5.226	5.138	5.064	5.002	4.948	4.901	4.859	4.823	4.790
32	496	.10	4.253	4.206	4.167	4.134	4.105	4.080	4.058	4.039	4.021	4.005
		.05	4.604	4.540	4.487	4.441	4.402	4.369	4.339	4.313	4.289	4.268
		.01	5.382	5.274	5.184	5.110	5.046	4.991	4.943	4.901	4.864	4.831
34	561	.10	4.294	4.246	4.207	4.173	4.144	4.119	4.096	4.076	4.058	4.042
		.05	4.647	4.582	4.528	4.482	4.442	4.408	4.377	4.351	4.327	4.306
		.01	5.428	5.318	5.228	5.152	5.087	5.031	4.982	4.940	4.902	4.868
36	630	.10	4.332	4.284	4.244	4.210	4.180	4.154	4.132	4.112	4.094	4.077
		.05	4.687	4.621	4.566	4.519	4.479	4.444	4.414	4.386	4.362	4.341
		.01	5.471	5.360	5.268	5.191	5.125	5.069	5.020	4.976	4.938	4.904
38	703	.10	4.368	4.320	4.279	4.244	4.214	4.188	4.165	4.145	4.126	4.110
		.05	4.724	4.658	4.602	4.555	4.514	4.479	4.448	4.420	4.396	4.374
		.01	5.512	5.400	5.307	5.228	5.162	5.104	5.054	5.011	4.972	4.937

Degrees of freedom v

k	$k*$	α	20	22	24	26	28	30	32	34	36	38
40	780	.10	4.402	4.353	4.312	4.277	4.247	4.220	4.197	4.176	4.157	4.141
		.05	4.760	4.692	4.636	4.588	4.547	4.511	4.480	4.452	4.427	4.405
		.01	5.551	5.437	5.343	5.264	5.196	5.138	5.087	5.043	5.004	4.968
42	861	.10	4.434	4.385	4.343	4.308	4.277	4.250	4.227	4.206	4.187	4.170
		.05	4.793	4.725	4.668	4.620	4.578	4.542	4.510	4.482	4.457	4.434
		.01	5.587	5.472	5.377	5.297	5.229	5.170	5.119	5.074	5.034	4.998
44	946	.10	4.464	4.415	4.373	4.337	4.306	4.279	4.255	4.234	4.215	4.198
		.05	4.825	4.756	4.698	4.650	4.607	4.571	4.539	4.510	4.485	4.462
		.01	5.622	5.505	5.409	5.329	5.260	5.200	5.149	5.103	5.063	5.027
46	1035	.10	4.493	4.443	4.401	4.364	4.333	4.306	4.282	4.260	4.241	4.224
		.05	4.855	4.786	4.727	4.678	4.636	4.599	4.566	4.537	4.512	4.489
		.01	5.655	5.537	5.440	5.359	5.289	5.229	5.177	5.131	5.090	5.054
48	1128	.10	4.521	4.470	4.427	4.391	4.359	4.332	4.307	4.286	4.266	4.249
		.05	4.884	4.814	4.755	4.705	4.662	4.625	4.592	4.563	4.537	4.514
		.01	5.686	5.568	5.470	5.387	5.317	5.257	5.204	5.157	5.116	5.080
50	1225	.10	4.547	4.496	4.453	4.416	4.384	4.356	4.332	4.310	4.290	4.273
		.05	4.911	4.841	4.781	4.731	4.688	4.650	4.617	4.588	4.562	4.538
		.01	5.716	5.597	5.498	5.415	5.344	5.283	5.230	5.183	5.141	5.104
60	1770	.10	4.663	4.610	4.565	4.527	4.494	4.465	4.440	4.417	4.397	4.378
		.05	5.033	4.959	4.898	4.846	4.801	4.762	4.727	4.697	4.670	4.645
		.01	5.849	5.725	5.623	5.537	5.463	5.400	5.344	5.296	5.252	5.214
70	2415	.10	4.759	4.705	4.659	4.619	4.585	4.556	4.529	4.506	4.485	4.466
		.05	5.133	5.058	4.995	4.941	4.895	4.854	4.819	4.787	4.759	4.734
		.01	5.960	5.833	5.727	5.639	5.563	5.497	5.440	5.390	5.345	5.305
80	3160	.10	4.841	4.786	4.738	4.698	4.663	4.633	4.606	4.582	4.560	4.541
		.05	5.219	5.142	5.078	5.023	4.975	4.934	4.897	4.865	4.836	4.810
		.01	6.055	5.925	5.817	5.726	5.648	5.581	5.522	5.470	5.424	5.383
100	4950	.10	4.976	4.918	4.869	4.827	4.791	4.759	4.731	4.706	4.684	4.664
		.05	5.361	5.281	5.214	5.157	5.107	5.064	5.026	4.993	4.962	4.935
		.01	6.211	6.076	5.964	5.870	5.789	5.719	5.658	5.604	5.556	5.513

TABLE M Critical values of the studentized maximum modulus distribution (*continued*)

Degrees of freedom v

k	$k*$	α	40	50	60	80	100	120	240	480	∞
20	190	.10	3.695	3.648	3.616	3.576	3.552	3.535	3.494	3.474	3.453
		.05	3.953	3.891	3.850	3.798	3.767	3.746	3.694	3.668	3.643
		.01	4.506	4.409	4.346	4.267	4.221	4.191	4.116	4.079	4.042
22	231	.10	3.756	3.708	3.675	3.633	3.608	3.591	3.549	3.527	3.505
		.05	4.015	3.950	3.907	3.854	3.821	3.800	3.746	3.719	3.693
		.01	4.567	4.467	4.401	4.320	4.273	4.241	4.164	4.126	4.088
24	276	.10	3.812	3.762	3.728	3.685	3.659	3.642	3.597	3.575	3.552
		.05	4.070	4.004	3.960	3.904	3.871	3.849	3.793	3.765	3.738
		.01	4.622	4.519	4.452	4.368	4.319	4.287	4.207	4.168	4.129
26	325	.10	3.862	3.811	3.776	3.732	3.705	3.687	3.642	3.618	3.595
		.05	4.120	4.053	4.007	3.950	3.916	3.893	3.836	3.807	3.778
		.01	4.672	4.567	4.497	4.412	4.362	4.328	4.247	4.206	4.167
28	378	.10	3.909	3.856	3.820	3.775	3.748	3.729	3.682	3.658	3.634
		.05	4.167	4.097	4.051	3.992	3.957	3.934	3.875	3.845	3.816
		.01	4.718	4.611	4.540	4.452	4.401	4.367	4.283	4.242	4.201
30	435	.10	3.951	3.898	3.861	3.815	3.787	3.768	3.720	3.695	3.670
		.05	4.210	4.139	4.091	4.031	3.995	3.971	3.911	3.881	3.850
		.01	4.761	4.651	4.579	4.490	4.437	4.402	4.316	4.274	4.233
32	496	.10	3.991	3.936	3.899	3.852	3.823	3.804	3.754	3.729	3.704
		.05	4.249	4.177	4.128	4.067	4.031	4.006	3.944	3.913	3.882
		.01	4.801	4.689	4.615	4.524	4.470	4.435	4.347	4.304	4.262
34	561	.10	4.028	3.972	3.934	3.886	3.857	3.837	3.786	3.761	3.735
		.05	4.286	4.213	4.163	4.101	4.064	4.038	3.975	3.944	3.912
		.01	4.838	4.724	4.649	4.557	4.502	4.466	4.376	4.333	4.289
36	630	.10	4.063	4.006	3.967	3.918	3.888	3.868	3.817	3.790	3.764
		.05	4.321	4.246	4.196	4.133	4.095	4.069	4.005	3.972	3.940
		.01	4.873	4.757	4.681	4.587	4.531	4.494	4.404	4.359	4.315
38	703	.10	4.095	4.038	3.999	3.949	3.918	3.898	3.845	3.818	3.791
		.05	4.354	4.278	4.227	4.163	4.124	4.098	4.032	3.999	3.966
		.01	4.906	4.788	4.711	4.616	4.559	4.521	4.429	4.384	4.339

Degrees of freedom v

k	k^*	α	40	50	60	80	100	120	240	480	∞
40	780	.10	4.126	4.067	4.028	3.977	3.946	3.925	3.872	3.845	3.817
		.05	4.385	4.308	4.256	4.191	4.151	4.125	4.058	4.025	3.991
		.01	4.937	4.818	4.739	4.643	4.585	4.547	4.453	4.407	4.362
42	861	.10	4.155	4.096	4.056	4.004	3.973	3.951	3.897	3.869	3.841
		.05	4.414	4.336	4.283	4.217	4.177	4.150	4.082	4.048	4.014
		.01	4.966	4.846	4.766	4.668	4.610	4.571	4.476	4.430	4.384
44	946	.10	4.182	4.122	4.082	4.030	3.998	3.976	3.921	3.893	3.864
		.05	4.441	4.363	4.309	4.242	4.202	4.174	4.106	4.071	4.037
		.01	4.994	4.872	4.792	4.692	4.633	4.594	4.498	4.451	4.404
46	1035	.10	4.208	4.148	4.107	4.054	4.022	4.000	3.944	3.915	3.886
		.05	4.468	4.388	4.334	4.266	4.225	4.197	4.128	4.093	4.058
		.01	5.021	4.898	4.816	4.716	4.656	4.616	4.519	4.471	4.424
48	1128	.10	4.233	4.172	4.131	4.077	4.044	4.022	3.966	3.937	3.907
		.05	4.493	4.412	4.358	4.289	4.247	4.219	4.149	4.113	4.078
		.01	5.047	4.922	4.840	4.738	4.677	4.637	4.538	4.490	4.442
50	1225	.10	4.257	4.195	4.153	4.099	4.066	4.044	3.986	3.957	3.927
		.05	4.517	4.435	4.381	4.311	4.269	4.240	4.169	4.133	4.097
		.01	5.071	4.945	4.862	4.759	4.698	4.657	4.557	4.508	4.460
60	1770	.10	4.362	4.297	4.254	4.197	4.162	4.139	4.078	4.046	4.015
		.05	4.623	4.538	4.480	4.407	4.363	4.333	4.258	4.220	4.181
		.01	5.179	5.047	4.960	4.852	4.788	4.746	4.641	4.589	4.538
70	2415	.10	4.449	4.382	4.337	4.278	4.242	4.217	4.154	4.121	4.087
		.05	4.711	4.623	4.564	4.488	4.441	4.410	4.331	4.291	4.251
		.01	5.269	5.133	5.043	4.931	4.864	4.819	4.710	4.656	4.603
80	3160	.10	4.523	4.455	4.408	4.347	4.310	4.284	4.218	4.184	4.149
		.05	4.787	4.696	4.635	4.556	4.509	4.476	4.394	4.353	4.311
		.01	5.347	5.206	5.113	4.998	4.929	4.883	4.769	4.714	4.659
100	4950	.10	4.645	4.574	4.525	4.461	4.422	4.395	4.325	4.288	4.251
		.05	4.911	4.816	4.752	4.670	4.619	4.585	4.498	4.454	4.409
		.01	5.474	5.328	5.230	5.108	5.036	4.987	4.868	4.809	4.750

TABLE N Shortest unbiased confidence limits for the mean from a Poisson distribution

This table furnishes the upper and lower 95% and 99% shortest unbiased confidence limits for the mean of a Poisson distribution. Limits are provided for observed frequencies in the range 0 to 100. Since the intervals do not restrict the division of probability between the two tails, they should not be used for one-sided confidence intervals.

For an observed count of $Y = 10$ in a sample, the 95% confidence limits, shown in the table, are 5.324 and 18.338. These represent the range of true values of the Poisson parameter that would not be rejected at the 5% level if we carried out a test of significance based on an observed count Y of 10.

For values of Y larger than 100 the following formulas can be used to provide a good approximation to the central (equal tails) confidence limits.

$$M_1 = Y - \frac{1}{2} + \frac{3}{8}z^2 - z\sqrt{Y - \frac{1}{2} + \frac{1}{8}z^2}$$

$$M_2 = Y + \frac{1}{2} + \frac{3}{8}z^2 + z\sqrt{Y + \frac{1}{2} + \frac{1}{8}z^2}$$

where z is the upper $\alpha/2$ value from a table of areas on the normal curve (such as Table A) and Y is the observed frequency. Use $z = 1.960$ or $z = 2.576$ for 95% or 99% confidence limits, respectively.

The use of this table is discussed in Box 7.4, Section 7.7.

The values in this table were computed using the δ_3 method described in E. L. Crow and R. S. Gardner (*Biometrika* **46**:441–453, 1959). This approach was used rather than the δ_4 method, which reduces the lengths of the confidence intervals for small values of Y at the expense of those for larger values of Y, because it is not clear why the lengths of the former should be considered more important than the latter. The confidence regions can sometimes be discontinuous. In such cases the tabled values give the extreme limits.

TABLE **N**   Shortest unbiased confidence limits for the mean from a Poisson distribution

| | Confidence coefficients | | | |
| | 0.95 | | 0.99 | |
Y	Lower	Upper	Lower	Upper
0	0.000	3.764	0.000	5.288
1	0.052	5.755	0.011	7.336
2	0.356	7.294	0.149	9.312
3	0.818	8.807	0.437	11.263
4	1.367	10.307	0.824	12.792
5	1.971	11.799	1.280	14.307
6	2.614	13.286	1.786	15.813
7	3.286	14.340	2.331	17.312
8	3.765	15.819	2.907	18.807
9	4.461	17.297	3.508	20.298
10	5.324	18.338	4.131	21.359
11	5.756	19.813	4.772	22.844
12	6.686	20.848	5.289	24.326
13	7.295	22.321	5.829	25.376
14	8.103	23.795	6.668	26.855
15	8.808	24.824	7.337	28.334
16	9.599	25.852	7.756	29.376
17	10.308	27.324	8.727	30.852
18	11.178	28.348	9.313	32.328
19	11.800	29.820	10.010	33.365
20	12.818	30.843	10.859	34.840
21	13.287	32.314	11.264	35.874
22	14.341	33.336	12.347	37.347
23	14.921	34.807	12.793	38.379
24	15.820	35.827	13.794	39.852
25	16.768	36.846	14.308	40.881
26	17.298	38.317	15.277	42.354
27	18.339	39.335	15.814	43.381
28	19.051	40.352	16.801	44.854
29	19.814	41.824	17.313	45.880
30	20.849	42.840	18.363	47.352
31	21.365	44.312	18.808	48.376
32	22.322	45.327	19.874	49.848
33	23.353	46.342	20.299	50.872
34	23.796	47.814	21.360	52.343
35	24.825	48.829	22.043	53.366
36	25.853	49.842	22.845	54.837
37	26.307	51.315	23.765	55.859
38	27.325	52.328	24.327	56.879
39	28.349	53.341	25.377	58.351
40	28.967	54.353	25.829	59.371
41	29.821	55.826	26.856	60.842
42	30.844	56.838	27.718	61.862
43	31.675	57.850	28.335	62.880
44	32.315	59.323	29.377	64.352
45	33.337	60.334	29.901	65.370
46	34.357	61.345	30.853	66.841
47	34.808	62.819	31.840	67.859
48	35.828	63.830	32.329	68.876
49	36.847	64.840	33.366	70.348
50	37.667	65.850	34.183	71.364

TABLE N   Shortest unbiased confidence limits for the mean from a Poisson distribution (*continued*)

| | **Confidence coefficients** | | | |
| | **0.95** | | **0.99** | |
Y	**Lower**	**Upper**	**Lower**	**Upper**
51	38.318	67.324	34.841	72.380
52	39.336	68.334	35.875	73.852
53	40.353	69.344	36.545	74.868
54	40.944	70.353	37.348	75.883
55	41.825	71.828	38.380	77.355
56	42.841	72.837	38.940	78.370
57	43.856	73.846	39.853	79.385
58	44.313	74.855	40.882	80.857
59	45.328	76.330	41.391	81.871
60	46.343	77.338	42.355	82.885
61	47.357	78.347	43.382	84.357
62	47.815	79.355	43.914	85.371
63	48.830	80.830	44.855	86.384
64	49.843	81.839	45.881	87.857
65	50.857	82.847	46.502	88.870
66	51.316	83.855	47.353	89.883
67	52.329	85.330	48.377	91.355
68	53.342	86.338	49.131	92.368
69	54.354	87.345	49.849	93.380
70	54.991	88.353	50.873	94.853
71	55.827	89.829	51.784	95.865
72	56.839	90.836	52.344	96.877
73	57.851	91.843	53.367	97.889
74	58.718	92.851	54.388	99.362
75	59.324	94.327	54.838	100.374
76	60.335	95.334	55.860	101.385
77	61.346	96.341	56.880	102.858
78	62.357	97.348	57.616	103.869
79	62.820	98.354	58.352	104.880
80	63.831	99.831	59.372	106.354
81	64.841	100.837	60.391	107.365
82	65.851	101.844	60.843	108.375
83	66.758	102.850	61.863	109.386
84	67.325	103.857	62.881	110.859
85	68.335	105.334	63.635	111.870
86	69.345	106.340	64.353	112.880
87	70.354	107.346	65.371	114.354
88	71.093	108.352	66.388	115.364
89	71.829	109.829	66.842	116.374
90	72.838	110.836	67.860	117.384
91	73.847	111.842	68.877	118.858
92	74.856	112.847	69.830	119.867
93	75.490	113.853	70.349	120.877
94	76.331	115.331	71.365	121.886
95	77.339	116.336	72.381	123.360
96	78.348	117.342	73.204	124.370
97	79.356	118.348	73.853	125.379
98	79.983	119.353	74.869	126.388
99	80.831	120.831	75.884	127.862
100	81.840	121.837	76.609	128.871

TABLE O Shortest unbiased confidence limits for the variance

This table furnishes special multiplication factors f_1 and f_2, which greatly simplify the computation of shortest unbiased confidence limits for the variance with confidence coefficients 0.95 and 0.99. The coefficients of the factors are provided for degrees of freedom v from 2 to 30 in increments of one and from $v = 40$ to 100 in increments of 10.

For a sample variance based on 10 items ($v = 9$) and a 0.95 confidence coefficient, the factors in the table are $f_1 = 0.4432$ and $f_2 = 3.048$; for a confidence coefficient of 0.99, the factors are $f_1 = 0.3585$ and $f_2 = 4.720$. For degrees of freedom > 100 use the ordinary (equal tails) method for computing confidence limits of a variance (Section 7.7). When the number of degrees of freedom is large, the two methods yield very similar results.

These factors are employed in the setting of confidence limits to variances (Section 7.7 and Box 7.3).

The factors in this table have been obtained by dividing the quantity $n - 1$ by the values in a table prepared by D. V. Lindley, D. A. East, and P. A. Hamilton (*Biometrika* **47**:433–437, 1960).

TABLE O   Shortest unbiased confidence limits for the variance

ν	Confidence coefficients		ν	Confidence coefficients		ν	Confidence coefficients	
	0.95	**0.99**		**0.95**	**0.99**		**0.95**	**0.99**
2	.2099	.1505	14	.5135	.4289	26	.6057	.5261
	23.605	114.489		2.354	3.244		1.825	2.262
3	.2681	.1983	15	.5242	.4399	27	.6110	.5319
	10.127	29.689		2.276	3.091		1.802	2.223
4	.3125	.2367	16	.5341	.4502	28	.6160	.5374
	6.590	15.154		2.208	2.961		1.782	2.187
5	.3480	.2685	17	.5433	.4598	29	.6209	.5427
	5.054	10.076		2.149	2.848		1.762	2.153
6	.3774	.2956	18	.5520	.4689	30	.6255	.5478
	4.211	7.637		2.097	2.750		1.744	2.122
7	.4025	.3192	19	.5601	.4774	40	.6636	.5900
	3.679	6.238		2.050	2.664		1.608	1.896
8	.4242	.3400	20	.5677	.4855	50	.6913	.6213
	3.314	5.341		2.008	2.588		1.523	1.760
9	.4432	.3585	21	.5749	.4931	60	.7128	.6458
	3.048	4.720		1.971	2.519		1.464	1.668
10	.4602	.3752	22	.5817	.5004	70	.7300	.6657
	2.844	4.265		1.936	2.458		1.421	1.607
11	.4755	.3904	23	.5882	.5073	80	.7443	.6824
	2.683	3.919		1.905	2.402		1.387	1.549
12	.4893	.4043	24	.5943	.5139	90	.7564	.6966
	2.553	3.646		1.876	2.351		1.360	1.508
13	.5019	.4171	25	.6001	.5201	100	.7669	.7090
	2.445	3.426		1.850	2.305		1.338	1.475

TABLE **P**   Shortest unbiased confidence limits for proportions

This table furnishes confidence limits for proportions based on the binomial distribution. The first part of the table gives limits for samples up to size $n = 30$. The arguments are n, sample size, and Y, number of items in the sample that exhibit a given property. Argument Y is tabled for integral values between 0 and n. For each sample size n and number of items Y with the given property, the 95% and 99% confidence limits for the proportion are shown. For example, for $Y = 8$ individuals showing the property out of a sample of $n = 20$, an incidence of the property of 0.4, the 95% confidence limits of this proportion are given as 0.2090 to 0.6277, and the 99% limits as 0.1635 to 0.6889.

The second part of the table is for larger samples ($n = 50$, 100, 200, 500, and 1,000, indicated across the top of the table). The arguments along the left margin of the table are now percentages from 0 to 50% in increments of 1%, rather than counts. The 95% and 99% confidence limits corresponding to a given percentage and sample size n are the functions given in two lines in the body of the table. For instance, the 99% confidence limits of an observed incidence of 12% in a sample of 500 are 0.0860 to 0.1621, as shown in the second of the two lines. Interpolation in this table between the furnished sample sizes can be achieved by means of the following formula for the lower limit:

$$L_1 = \frac{L_1^- n^- (n^+ - n) + L_1^+ n^+ (n - n^-)}{n(n^+ - n^-)}$$

where n is the size of the observed sample, n^- and n^+ are the next lower and upper tabled sample sizes, respectively, L_1^- and L_1^+ are corresponding tabled confidence limits for these sample sizes, and L_1 is the lower confidence limit to be found by interpolation. The upper confidence limit, L_2, can be obtained by a corresponding formula by substituting 2 for the subscript 1.

Consider the following example, in which we set 95% confidence limits to an observed percentage of 25% in a sample size of 80. The tabled 95% limits for $n = 50$ are $0.1462 - 0.3891$. For $n = 100$ the corresponding tabled limits are $0.1741 - 0.3444$. When we substitute the values for the lower limits in the formula shown above, we obtain

$$L_1 = \frac{(0.1462)(50)(100 - 80) + (0.1741)(100)(80 - 50)}{80(100 - 50)} = 0.1671$$

for the lower confidence limit and

$$L_2 = \frac{(0.3891)(50)(100 - 80) + (0.3444)(100)(80 - 50)}{80(100 - 50)} = 0.3556$$

for the upper confidence limit.

The tabled values in italics for sample size 50 are limits for percentages that could not be obtained in any real sampling problem (for example, 25% in 50 items) but are necessary for interpolation. For percentages greater than 50%, look up the complementary percentage as the argument. The complements of the tabled binomial confidence limits are the desired limits.

Confidence limits for percentages are used in many different applications (for an example see Box 7.4) and in a variety of problems with frequencies (Chapters 17 and 18).

The values in this table were computed using the Stern method δ_3 as described in E. L. Crow (*Biometrika* **43**:423–435, 1956). This approach was used rather than the δ_4 method advocated by Crow. In the δ_4 method the lengths of the confidence intervals for small values of Y are reduced at the expense of those for larger values of Y. It is not clear why the lengths of the former should be considered more important than the latter. The confidence regions for the binomial proportion can sometimes be discontinuous. In such cases the tabled values give the extreme limits.

TABLE P   Shortest unbiased confidence limits for proportions

| | | Confidence coefficients | | | |
| | | 0.95 | | 0.99 | |
n	Y	Lower	Upper	Lower	Upper
1	0	.0000	.9500	.0000	.9900
	1	.0500	1.0000	.0100	1.0000
2	0	.0000	.7763	.0000	.9000
	1	.0254	.9746	.0051	.9949
	2	.2237	1.0000	.1000	1.0000
3	0	.0000	.6315	.0000	.7845
	1	.0170	.8646	.0034	.9410
	2	.1354	.9830	.0590	.9966
	3	.3685	1.0000	.2155	1.0000
4	0	.0000	.5271	.0000	.6837
	1	.0128	.7513	.0026	.8591
	2	.0977	.9023	.0420	.9580
	3	.2487	.9872	.1409	.9974
	4	.4729	1.0000	.3163	1.0000
5	0	.0000	.5000	.0000	.6018
	1	.0103	.6574	.0021	.7779
	2	.0765	.8107	.0327	.8943
	3	.1893	.9235	.1057	.9673
	4	.3426	.9897	.2221	.9979
	5	.5000	1.0000	.3982	1.0000
6	0	.0000	.4113	.0000	.5358
	1	.0086	.5886	.0017	.7056
	2	.0629	.7286	.0268	.8269
	3	.1532	.8468	.0848	.9152
	4	.2714	.9371	.1731	.9732
	5	.4114	.9914	.2944	.9983
	6	.5887	1.0000	.4642	1.0000
7	0	.0000	.3771	.0000	.5000
	1	.0074	.5542	.0015	.6433
	2	.0534	.6587	.0227	.7636
	3	.1288	.7746	.0709	.8577
	4	.2254	.8712	.1423	.9291
	5	.3413	.9466	.2364	.9773
	6	.4458	.9926	.3567	.9985
	7	.6229	1.0000	.5000	1.0000
8	0	.0000	.3646	.0000	.4506
	1	.0064	.5000	.0013	.5899
	2	.0464	.6353	.0197	.7067
	3	.1112	.7107	.0609	.8017
	4	.1930	.8070	.1210	.8790
	5	.2893	.8888	.1983	.9391
	6	.3647	.9536	.2933	.9803
	7	.5000	.9936	.4101	.9987
	8	.6354	1.0000	.5494	1.0000
9	0	.0000	.3233	.0000	.4317
	1	.0057	.4434	.0012	.5682
	2	.0411	.5582	.0174	.6563
	3	.0978	.6766	.0534	.7499

TABLE P Shortest unbiased confidence limits for proportions
(*continued*)

		Confidence coefficients			
		0.95		0.99	
n	*Y*	Lower	Upper	Lower	Upper
	4	.1688	.7486	.1053	.8290
	5	.2514	.8312	.1710	.8947
	6	.3234	.9022	.2501	.9466
	7	.4418	.9589	.3437	.9826
	8	.5566	.9943	.4318	.9988
	9	.6767	1.0000	.5683	1.0000
10	0	.0000	.2908	.0000	.3832
	1	.0052	.4464	.0011	.5123
	2	.0368	.5535	.0156	.6167
	3	.0873	.6194	.0476	.7028
	4	.1501	.7091	.0933	.7816
	5	.2225	.7775	.1505	.8495
	6	.2909	.8499	.2184	.9067
	7	.3806	.9127	.2972	.9524
	8	.4465	.9632	.3833	.9844
	9	.5536	.9948	.4877	.9989
	10	.7092	1.0000	.6168	1.0000
11	0	.0000	.2645	.0000	.3591
	1	.0047	.4044	.0010	.5000
	2	.0334	.5000	.0141	.5927
	3	.0789	.5955	.0429	.6604
	4	.1351	.6671	.0837	.7377
	5	.1996	.7354	.1344	.8060
	6	.2646	.8004	.1940	.8656
	7	.3329	.8649	.2623	.9163
	8	.4045	.9211	.3396	.9571
	9	.5000	.9666	.4073	.9859
	10	.5956	.9953	.5000	.9990
	11	.7355	1.0000	.6409	1.0000
12	0	.0000	.2426	.0000	.3545
	1	.0043	.3701	.0009	.4527
	2	.0305	.4571	.0129	.5472
	3	.0719	.5428	.0390	.6454
	4	.1229	.6298	.0759	.6975
	5	.1811	.7060	.1215	.7651
	6	.2427	.7573	.1747	.8253
	7	.2940	.8189	.2349	.8785
	8	.3702	.8771	.3025	.9241
	9	.4572	.9281	.3546	.9610
	10	.5429	.9695	.4528	.9871
	11	.6299	.9957	.5473	.9991
	12	.7574	1.0000	.6455	1.0000
13	0	.0000	.2251	.0000	.3251
	1	.0040	.3415	.0008	.4289
	2	.0281	.4339	.0119	.5233
	3	.0661	.5196	.0358	.5937
	4	.1127	.5865	.0695	.6748
	5	.1657	.6584	.1109	.7271
	6	.2240	.7395	.1589	.7871
	7	.2605	.7760	.2129	.8411
	8	.3416	.8343	.2729	.8891
	9	.4135	.8873	.3252	.9305

TABLE P Shortest unbiased confidence limits for proportions
(*continued*)

| | | Confidence coefficients | | | |
| | | 0.95 | | 0.99 | |
n	*Y*	Lower	Upper	Lower	Upper
	10	.4804	.9339	.4063	.9642
	11	.5661	.9719	.4767	.9881
	12	.6585	.9960	.5711	.9992
	13	.7749	1.0000	.6749	1.0000
14	0	.0000	.2381	.0000	.3005
	1	.0037	.3171	.0008	.4192
	2	.0260	.4256	.0110	.5000
	3	.0612	.5000	.0331	.5807
	4	.1041	.5743	.0641	.6364
	5	.1528	.6289	.1020	.6994
	6	.2061	.6828	.1457	.7512
	7	.2382	.7618	.1948	.8052
	8	.3172	.7939	.2488	.8543
	9	.3711	.8472	.3006	.8980
	10	.4257	.8959	.3636	.9359
	11	.5000	.9388	.4193	.9669
	12	.5744	.9740	.5000	.9890
	13	.6829	.9963	.5808	.9992
	14	.7619	1.0000	.6995	1.0000
15	0	.0000	.2222	.0000	.2795
	1	.0035	.3020	.0007	.3891
	2	.0243	.3967	.0102	.4634
	3	.0569	.4657	.0308	.5365
	4	.0967	.5342	.0594	.6108
	5	.1417	.6032	.0944	.6724
	6	.1909	.6676	.1346	.7204
	7	.2223	.7060	.1795	.7712
	8	.2940	.7777	.2288	.8205
	9	.3324	.8091	.2796	.8654
	10	.3968	.8583	.3276	.9056
	11	.4658	.9033	.3892	.9406
	12	.5343	.9431	.4635	.9692
	13	.6033	.9757	.5366	.9898
	14	.6980	.9965	.6109	.9993
	15	.7778	1.0000	.7205	1.0000
16	0	.0000	.2083	.0000	.2638
	1	.0033	.3054	.0007	.3633
	2	.0227	.3716	.0096	.4513
	3	.0532	.4361	.0287	.5245
	4	.0903	.5000	.0554	.5791
	5	.1322	.5638	.0879	.6366
	6	.1778	.6283	.1251	.7046
	7	.2084	.6945	.1665	.7393
	8	.2720	.7280	.2118	.7882
	9	.3055	.7916	.2607	.8335
	10	.3717	.8222	.2954	.8749
	11	.4362	.8678	.3634	.9121
	12	.5000	.9097	.4209	.9446
	13	.5639	.9468	.4755	.9713
	14	.6284	.9773	.5487	.9904
	15	.6946	.9967	.6367	.9993
	16	.7917	1.0000	.7362	1.0000

TABLE P Shortest unbiased confidence limits for proportions (*continued*)

n	Y	Confidence coefficients 0.95 Lower	Upper	0.99 Lower	Upper
17	0	.0000	.1961	.0000	.2713
	1	.0031	.2873	.0006	.3462
	2	.0214	.3497	.0090	.4353
	3	.0499	.4165	.0270	.5000
	4	.0847	.4887	.0519	.5646
	5	.1238	.5441	.0822	.6195
	6	.1664	.5937	.1169	.6615
	7	.1962	.6502	.1553	.7286
	8	.2531	.7126	.1972	.7577
	9	.2874	.7469	.2423	.8028
	10	.3498	.8038	.2714	.8447
	11	.4063	.8336	.3385	.8831
	12	.4559	.8762	.3805	.9178
	13	.5113	.9153	.4354	.9481
	14	.5835	.9501	.5000	.9730
	15	.6503	.9786	.5647	.9910
	16	.7127	.9969	.6538	.9994
	17	.8039	1.0000	.7287	1.0000
18	0	.0000	.1852	.0000	.2556
	1	.0029	.2713	.0006	.3473
	2	.0202	.3302	.0085	.4095
	3	.0471	.4140	.0254	.4699
	4	.0797	.4714	.0488	.5300
	5	.1165	.5285	.0772	.5904
	6	.1564	.5859	.1096	.6526
	7	.1853	.6253	.1455	.6860
	8	.2365	.6697	.1845	.7443
	9	.2714	.7286	.2263	.7737
	10	.3303	.7635	.2557	.8155
	11	.3747	.8147	.3140	.8545
	12	.4141	.8436	.3474	.8904
	13	.4715	.8835	.4096	.9228
	14	.5286	.9203	.4700	.9512
	15	.5860	.9529	.5301	.9746
	16	.6698	.9798	.5905	.9915
	17	.7287	.9971	.6527	.9994
	18	.8148	1.0000	.7444	1.00000
19	0	.0000	.1755	.0000	.2417
	1	.0027	.2570	.0006	.3281
	2	.0191	.3157	.0081	.3868
	3	.0445	.3919	.0240	.4551
	4	.0753	.4461	.0461	.5152
	5	.1100	.5000	.0728	.5643
	6	.1475	.5538	.1033	.6131
	7	.1756	.6080	.1369	.6718
	8	.2219	.6551	.1733	.7074
	9	.2571	.6882	.2124	.7582
	10	.3118	.7429	.2418	.7876
	11	.3449	.7781	.2926	.8267
	12	.3920	.8244	.3282	.8631
	13	.4462	.8525	.3869	.8967
	14	.5000	.8900	.4357	.9272
	15	.5539	.9247	.4848	.9539
	16	.6081	.9555	.5449	.9760

TABLE P Shortest unbiased confidence limits for proportions (*continued*)

n	Y	Confidence coefficients 0.95		0.99	
		Lower	Upper	Lower	Upper
	17	.6843	.9809	.6132	.9919
	18	.7430	.9973	.6719	.9994
	19	.8245	1.0000	.7583	1.0000
20	0	.0000	.1668	.0000	.2292
	1	.0026	.2442	.0006	.3110
	2	.0181	.3199	.0076	.3747
	3	.0422	.3722	.0228	.4458
	4	.0714	.4235	.0437	.5000
	5	.1041	.4745	.0689	.5541
	6	.1396	.5254	.0976	.6010
	7	.1669	.5764	.1292	.6372
	8	.2090	.6277	.1635	.6889
	9	.2443	.6800	.2001	.7261
	10	.2928	.7072	.2293	.7707
	11	.3200	.7557	.2739	.7999
	12	.3723	.7910	.3111	.8365
	13	.4236	.8331	.3628	.8708
	14	.4746	.8604	.3990	.9024
	15	.5255	.8959	.4459	.9311
	16	.5765	.9286	.5000	.9563
	17	.6278	.9578	.5542	.9772
	18	.6801	.9819	.6253	.9924
	19	.7558	.9974	.6890	.9994
	20	.8332	1.0000	.7708	1.0000
21	0	.0000	.1589	.0000	.2180
	1	.0025	.2326	.0005	.2956
	2	.0172	.3046	.0073	.3713
	3	.0401	.3543	.0216	.4233
	4	.0679	.4032	.0415	.4745
	5	.0989	.4552	.0654	.5254
	6	.1325	.5060	.0925	.5766
	7	.1590	.5510	.1224	.6286
	8	.1974	.5967	.1547	.6607
	9	.2327	.6456	.1892	.7043
	10	.2758	.6953	.2181	.7428
	11	.3047	.7242	.2572	.7819
	12	.3544	.7673	.2957	.8108
	13	.4033	.8026	.3393	.8453
	14	.4490	.8410	.3714	.8776
	15	.4940	.8675	.4234	.9075
	16	.5448	.9011	.4746	.9346
	17	.5968	.9321	.5255	.9585
	18	.6457	.9599	.5767	.9784
	19	.6954	.9828	.6287	.9927
	20	.7674	.9975	.7044	.9995
	21	.8411	1.0000	.7820	1.0000
22	0	.0000	.1517	.0000	.2078
	1	.0024	.2221	.0005	.2818
	2	.0164	.2907	.0069	.3537
	3	.0383	.3382	.0206	.4031
	4	.0646	.3889	.0395	.4536
	5	.0942	.4534	.0622	.5045
	6	.1261	.5000	.0879	.5497

TABLE P Shortest unbiased confidence limits for proportions (*continued*)

		Confidence coefficients			
		0.95		0.99	
n	*Y*	Lower	Upper	Lower	Upper
	7	.1518	.5465	.1163	.5968
	8	.1870	.5823	.1468	.6462
	9	.2222	.6174	.1794	.6819
	10	.2605	.6617	.2079	.7181
	11	.2908	.7092	.2422	.7578
	12	.3383	.7395	.2819	.7921
	13	.3826	.7778	.3181	.8206
	14	.4177	.8130	.3538	.8532
	15	.4535	.8482	.4032	.8837
	16	.5000	.8739	.4503	.9121
	17	.5466	.9058	.4955	.9378
	18	.6111	.9354	.5464	.9605
	19	.6618	.9617	.5969	.9794
	20	.7093	.9836	.6463	.9931
	21	.7779	.9976	.7182	.9995
	22	.8483	1.0000	.7922	1.0000
23	0	.0000	.1451	.0000	.1986
	1	.0023	.2125	.0005	.2692
	2	.0157	.2781	.0066	.3377
	3	.0366	.3235	.0197	.3864
	4	.0617	.3887	.0377	.4533
	5	.0899	.4334	.0593	.5000
	6	.1203	.4778	.0838	.5466
	7	.1452	.5221	.1107	.5801
	8	.1777	.5665	.1397	.6159
	9	.2126	.6112	.1706	.6622
	10	.2466	.6398	.1987	.7016
	11	.2782	.6764	.2287	.7307
	12	.3236	.7218	.2693	.7713
	13	.3602	.7534	.2984	.8013
	14	.3888	.7874	.3378	.8294
	15	.4335	.8223	.3841	.8603
	16	.4779	.8548	.4199	.8893
	17	.5222	.8797	.4534	.9162
	18	.5666	.9101	.5000	.9407
	19	.6113	.9383	.5467	.9623
	20	.6765	.9634	.6136	.9803
	21	.7219	.9843	.6623	.9934
	22	.7875	.9977	.7308	.9995
	23	.8549	1.0000	.8014	1.0000
24	0	.0000	.1391	.0000	.1901
	1	.0022	.2037	.0005	.2589
	2	.0151	.2665	.0064	.3232
	3	.0350	.3100	.0188	.3885
	4	.0591	.3724	.0360	.4334
	5	.0859	.4151	.0567	.4778
	6	.1150	.4576	.0800	.5221
	7	.1392	.5000	.1057	.5665
	8	.1691	.5423	.1333	.6114
	9	.2038	.5848	.1626	.6379
	10	.2340	.6275	.1902	.6767
	11	.2666	.6611	.2164	.7204
	12	.3101	.6899	.2574	.7426
	13	.3389	.7334	.2796	.7836

TABLE P Shortest unbiased confidence limits for proportions (*continued*)

		Confidence coefficients			
		0.95		**0.99**	
n	*Y*	**Lower**	**Upper**	**Lower**	**Upper**
	14	.3725	.7660	.3233	.8098
	15	.4152	.7962	.3621	.8374
	16	.4577	.8309	.3886	.8667
	17	.5000	.8608	.4335	.8943
	18	.5424	.8850	.4779	.9200
	19	.5849	.9141	.5222	.9433
	20	.6276	.9409	.5666	.9640
	21	.6900	.9650	.6115	.9812
	22	.7335	.9849	.6768	.9936
	23	.7963	.9978	.7411	.9995
	24	.8609	.9578	1.0000	1.0000
25	0	.0000	.1336	.0000	.1824
	1	.0021	.1956	.0005	.2649
	2	.0145	.2559	.0061	.3099
	3	.0336	.3031	.0181	.3723
	4	.0566	.3574	.0345	.4152
	5	.0823	.3984	.0543	.4577
	6	.1101	.4391	.0766	.5000
	7	.1337	.4797	.1011	.5422
	8	.1614	.5202	.1274	.5847
	9	.1957	.5608	.1553	.6276
	10	.2224	.6015	.1825	.6583
	11	.2560	.6425	.2052	.6900
	12	.2959	.6829	.2452	.7350
	13	.3171	.7041	.2650	.7548
	14	.3575	.7440	.3100	.7948
	15	.3985	.7776	.3417	.8175
	16	.4392	.8043	.3724	.8447
	17	.4798	.8386	.4153	.8726
	18	.5203	.8663	.4578	.8989
	19	.5609	.8899	.5000	.9234
	20	.6016	.9177	.5423	.9457
	21	.6426	.9434	.5848	.9655
	22	.6969	.9664	.6277	.9819
	23	.7441	.9855	.6901	.9939
	24	.8044	.9979	.7351	.9995
	25	.8664	1.0000	.8176	1.0000
26	0	.0000	.1285	.0000	.1752
	1	.0020	.1881	.0004	.2544
	2	.0139	.2460	.0059	.2976
	3	.0322	.3037	.0174	.3574
	4	.0544	.3436	.0331	.3986
	5	.0790	.3830	.0521	.4416
	6	.1056	.4221	.0734	.4865
	7	.1286	.4650	.0969	.5262
	8	.1542	.5055	.1220	.5623
	9	.1882	.5421	.1487	.6013
	10	.2117	.5778	.1753	.6425
	11	.2461	.6169	.1947	.6778
	12	.2821	.6563	.2340	.7023
	13	.3038	.6962	.2545	.7455
	14	.3437	.7179	.2977	.7660
	15	.3831	.7539	.3222	.8053
	16	.4222	.7883	.3575	.8247

TABLE P Shortest unbiased confidence limits for proportions (*continued*)

		Confidence coefficients			
		0.95		**0.99**	
n	*Y*	**Lower**	**Upper**	**Lower**	**Upper**
	17	.4579	.8118	.3987	.8513
	18	.4945	.8458	.4377	.8780
	19	.5350	.8714	.4738	.9031
	20	.5779	.8944	.5135	.9266
	21	.6170	.9210	.5584	.9479
	22	.6564	.9456	.6014	.9669
	23	.6963	.9678	.6426	.9826
	24	.7540	.9861	.7024	.9941
	25	.8119	.9980	.7456	.9996
	26	.8715	1.0000	.8248	1.0000
27	0	.0000	.1238	.0000	.1687
	1	.0019	.1812	.0004	.2448
	2	.0134	.2370	.0056	.3033
	3	.0310	.2924	.0167	.3438
	4	.0523	.3308	.0319	.3844
	5	.0760	.3688	.0500	.4410
	6	.1015	.4147	.0705	.4803
	7	.1239	.4622	.0930	.5196
	8	.1477	.5000	.1171	.5589
	9	.1813	.5377	.1426	.5871
	10	.2018	.5704	.1688	.6172
	11	.2371	.5981	.1849	.6561
	12	.2694	.6311	.2238	.6966
	13	.2925	.6691	.2449	.7159
	14	.3309	.7075	.2841	.7551
	15	.3689	.7306	.3034	.7762
	16	.4019	.7629	.3439	.8151
	17	.4296	.7982	.3828	.8312
	18	.4623	.8187	.4129	.8574
	19	.5000	.8523	.4411	.8829
	20	.5378	.8761	.4804	.9070
	21	.5853	.8985	.5197	.9295
	22	.6312	.9240	.5590	.9500
	23	.6692	.9477	.6156	.9681
	24	.7076	.9690	.6562	.9833
	25	.7630	.9866	.6967	.9944
	26	.8188	.9981	.7552	.9996
	27	.8762	1.0000	.8313	1.0000
28	0	.0000	.1194	.0000	.1772
	1	.0019	.1748	.0004	.2358
	2	.0129	.2285	.0054	.2921
	3	.0299	.2820	.0161	.3311
	4	.0504	.3190	.0307	.3865
	5	.0732	.3571	.0481	.4246
	6	.0977	.4091	.0679	.4623
	7	.1195	.4455	.0894	.5000
	8	.1417	.4818	.1125	.5376
	9	.1749	.5181	.1370	.5753
	10	.1925	.5544	.1627	.6134
	11	.2286	.5908	.1773	.6359
	12	.2576	.6194	.2144	.6688
	13	.2821	.6451	.2359	.7078
	14	.3191	.6809	.2716	.7284
	15	.3549	.7179	.2922	.7641

TABLE P Shortest unbiased confidence limits for proportions (*continued*)

n	Y	Confidence coefficients 0.95		0.99	
		Lower	**Upper**	**Lower**	**Upper**
	16	.3806	.7424	.3312	.7856
	17	.4092	.7714	.3641	.8227
	18	.4456	.8075	.3866	.8373
	19	.4819	.8251	.4247	.8630
	20	.5182	.8583	.4624	.8875
	21	.5545	.8805	.5000	.9106
	22	.5909	.9023	.5377	.9321
	23	.6429	.9268	.5754	.9519
	24	.6810	.9496	.6135	.9693
	25	.7180	.9701	.6689	.9839
	26	.7715	.9871	.7079	.9946
	27	.8252	.9981	.7642	.9996
	28	.8806	1.0000	.8228	1.0000
29	0	.0000	.1153	.0000	.1710
	1	.0018	.1688	.0004	.2275
	2	.0124	.2207	.0052	.2818
	3	.0288	.2723	.0155	.3194
	4	.0486	.3080	.0296	.3727
	5	.0705	.3595	.0464	.4094
	6	.0942	.3949	.0654	.4457
	7	.1154	.4300	.0861	.4819
	8	.1361	.4650	.1083	.5180
	9	.1689	.5000	.1318	.5542
	10	.1837	.5349	.1566	.5905
	11	.2208	.5699	.1711	.6272
	12	.2466	.6050	.2058	.6536
	13	.2724	.6404	.2276	.6805
	14	.3081	.6608	.2601	.7181
	15	.3392	.6919	.2819	.7399
	16	.3596	.7276	.3195	.7724
	17	.3950	.7534	.3464	.7942
	18	.4301	.7792	.3728	.8289
	19	.4651	.8163	.4095	.8434
	20	.5000	.8311	.4458	.8682
	21	.5350	.8639	.4820	.8917
	22	.5700	.8846	.5181	.9139
	23	.6051	.9058	.5543	.9346
	24	.6405	.9295	.5906	.9536
	25	.6920	.9514	.6273	.9704
	26	.7277	.9712	.6806	.9845
	27	.7793	.9876	.7182	.9948
	28	.8312	.9982	.7725	.9996
	29	.8847	1.0000	.8290	1.0000
30	0	.0000	.1115	.0000	.1652
	1	.0018	.1772	.0004	.2198
	2	.0120	.2134	.0051	.2722
	3	.0279	.2632	.0150	.3102
	4	.0469	.2978	.0285	.3599
	5	.0681	.3475	.0448	.3953
	6	.0909	.3816	.0631	.4303
	7	.1116	.4156	.0830	.4690
	8	.1309	.4494	.1044	.5051
	9	.1633	.4831	.1270	.5379
	10	.1773	.5168	.1508	.5696

TABLE P Shortest unbiased confidence limits for proportions
(*continued*)

		Confidence coefficients			
		0.95		0.99	
n	*Y*	Lower	Upper	Lower	Upper
	11	.2135	.5505	.1653	.6046
	12	.2363	.5843	.1978	.6400
	13	.2633	.6183	.2199	.6712
	14	.2979	.6524	.2494	.6920
	15	.3242	.6758	.2723	.7277
	16	.3476	.7021	.3080	.7506
	17	.3817	.7367	.3288	.7801
	18	.4157	.7637	.3600	.8022
	19	.4495	.7865	.3954	.8347
	20	.4832	.8227	.4304	.8492
	21	.5169	.8367	.4621	.8730
	22	.5506	.8691	.4949	.8956
	23	.5844	.8884	.5310	.9170
	24	.6184	.9091	.5697	.9369
	25	.6525	.9319	.6047	.9552
	26	.7022	.9531	.6401	.9715
	27	.7368	.9721	.6898	.9850
	28	.7866	.9880	.7278	.9949
	29	.8228	.9982	.7802	.9996
	30	.8885	1.0000	.8348	1.0000

TABLE P Shortest unbiased confidence limits for proportions (*continued*)

		n								

		50		100		200		500		1000	
%	P	L	U	L	U	L	U	L	U	L	U
0	.95	.0000	.0749	.0000	.0375	.0000	.0187	.0000	.0075	.0000	.0037
	.99	.0000	.0985	.0000	.0530	.0000	.0264	.0000	.0105	.0000	.0052
1	.95	*.0006*	*.0907*	.0006	.0533	.0018	.0364	.0040	.0235	.0054	.0183
	.99	*.0001*	*.1190*	.0002	.0735	.0008	.0466	.0026	.0277	.0042	.0213
2	.95	.0011	.1065	.0036	.0727	.0069	.0514	.0107	.0366	.0129	.0308
	.99	.0003	.1394	.0015	.0892	.0042	.0619	.0084	.0427	.0109	.0344
3	.95	*.0042*	*.1217*	.0083	.0836	.0132	.0641	.0177	.0487	.0209	.0428
	.99	*.0016*	*.1551*	.0044	.1046	.0090	.0770	.0147	.0558	.0184	.0469
4	.95	.0072	.1369	.0138	.0985	.0188	.0768	.0257	.0607	.0294	.0543
	.99	.0030	.1708	.0084	.1199	.0147	.0898	.0218	.0688	.0264	.0589
5	.95	*.0119*	*.1519*	.0200	.1134	.0268	.0893	.0338	.0727	.0379	.0653
	.99	*.0060*	*.1862*	.0130	.1350	.0210	.1047	.0288	.0809	.0345	.0704
6	.95	.0166	.1669	.0265	.1238	.0341	.1018	.0417	.0847	.0464	.0763
	.99	.0089	.2016	.0182	.1500	.0265	.1174	.0368	.0929	.0430	.0819
7	.95	*.0222*	*.1774*	.0334	.1386	.0415	.1143	.0500	.0957	.0554	.0878
	.99	*.0129*	*.2168*	.0238	.1606	.0341	.1300	.0449	.1049	.0515	.0934
8	.95	.0278	.1879	.0376	.1534	.0494	.1268	.0588	.1077	.0644	.0983
	.99	.0169	.2321	.0297	.1756	.0403	.1426	.0529	.1160	.0600	.1045
9	.95	*.0340*	*.2027*	.0460	.1637	.0568	.1392	.0677	.1187	.0734	.1093
	.99	*.0216*	*.2428*	.0359	.1906	.0467	.1551	.0609	.1280	.0685	.1160
10	.95	.0403	.2176	.0534	.1740	.0642	.1494	.0757	.1297	.0824	.1203
	.99	.0264	.2535	.0424	.2010	.0544	.1677	.0689	.1391	.0775	.1270
11	.95	*.0470*	*.2279*	.0588	.1887	.0716	.1618	.0848	.1407	.0919	.1308
	.99	*.0317*	*.2687*	.0491	.2159	.0620	.1779	.0780	.1510	.0865	.1380
12	.95	.0536	.2383	.0687	.1989	.0806	.1720	.0938	.1517	.1009	.1418
	.99	.0370	.2838	.0531	.2263	.0696	.1905	.0860	.1621	.0956	.1485
13	.95	*.0604*	*.2531*	.0728	.2091	.0894	.1844	.1028	.1627	.1104	.1523
	.99	*.0427*	*.2943*	.0614	.2366	.0771	.2030	.0950	.1731	.1046	.1595
14	.95	.0672	.2679	.0837	.2239	.0968	.1945	.1118	.1737	.1194	.1629
	.99	.0485	.3048	.0695	.2515	.0846	.2132	.1036	.1842	.1136	.1705
15	.95	*.0711*	*.2782*	.0890	.2340	.1060	.2069	.1208	.1838	.1289	.1734
	.99	*.0546*	*.3200*	.0736	.2618	.0933	.2257	.1120	.1952	.1226	.1811
16	.95	.0750	.2884	.0986	.2442	.1144	.2170	.1298	.1948	.1384	.1843
	.99	.0607	.3351	.0831	.2721	.1023	.2358	.1211	.2062	.1316	.1921
17	.95	*.0848*	*.2985*	.1066	.2590	.1218	.2295	.1388	.2058	.1479	.1948
	.99	*.0672*	*.3455*	.0893	.2870	.1101	.2483	.1301	.2172	.1411	.2026
18	.95	.0946	.3087	.1135	.2691	.1320	.2395	.1478	.2168	.1569	.2054
	.99	.0736	.3559	.0971	.2973	.1175	.2585	.1392	.2283	.1501	.2131
19	.95	*.1006*	*.3236*	.1239	.2792	.1393	.2520	.1579	.2268	.1664	.2154
	.99	*.0804*	*.3663*	.1047	.3075	.1259	.2710	.1482	.2393	.1596	.2236
20	.95	.1066	.3385	.1287	.2893	.1495	.2620	.1668	.2378	.1759	.2259
	.99	.0871	.3767	.1120	.3177	.1353	.2811	.1572	.2493	.1690	.2341
21	.95	*.1151*	*.3487*	.1387	.2994	.1569	.2721	.1758	.2478	.1854	.2364
	.99	*.0928*	*.3870*	.1200	.3327	.1427	.2936	.1662	.2603	.1781	.2451
22	.95	.1235	.3588	.1490	.3142	.1670	.2821	.1851	.2588	.1949	.2469
	.99	.0986	.3973	.1278	.3429	.1516	.3037	.1752	.2714	.1877	.2556
23	.95	*.1303*	*.3689*	.1535	.3243	.1751	.2946	.1949	.2688	.2047	.2574
	.99	*.1045*	*.4076*	.1351	.3530	.1604	.3138	.1843	.2814	.1972	.2657
24	.95	.1370	.3790	.1638	.3344	.1845	.3046	.2039	.2798	.2145	.2674
	.99	.1104	.4179	.1445	.3632	.1678	.3263	.1933	.2924	.2067	.2762
25	.95	*.1462*	*.3891*	.1741	.3444	.1946	.3147	.2139	.2898	.2240	.2779
	.99	*.1194*	*.4298*	.1501	.3734	.1780	.3364	.2023	.3025	.2160	.2867

TABLE P Shortest unbiased confidence limits for proportions (*continued*)

%	P	50 L	50 U	100 L	100 U	200 L	200 U	500 L	500 U	1000 L	1000 U
26	.95	.1555	.3992	.1807	.3545	.2020	.3247	.2229	.3008	.2335	.2884
	.99	.1284	.4417	.1607	.3853	.1854	.3465	.2118	.3135	.2253	.2972
27	.95	.1613	.4093	.1888	.3645	.2121	.3372	.2324	.3108	.2430	.2984
	.99	.1340	.4554	.1668	.3986	.1957	.3566	.2214	.3235	.2348	.3077
28	.95	.1670	.4194	.1990	.3755	.2214	.3472	.2419	.3208	.2530	.3089
	.99	.1395	.4691	.1757	.4087	.2031	.3691	.2304	.3345	.2446	.3177
29	.95	.1775	.4295	.2092	.3895	.2296	.3572	.2514	.3318	.2625	.3189
	.99	.1482	.4794	.1859	.4189	.2133	.3792	.2394	.3446	.2543	.3282
30	.95	.1880	.4396	.2142	.3996	.2396	.3672	.2609	.3419	.2720	.3294
	.99	.1568	.4897	.1907	.4290	.2207	.3892	.2494	.3556	.2638	.3387
31	.95	.1946	.4496	.2240	.4096	.2495	.3773	.2709	.3519	.2820	.3394
	.99	.1638	.5000	.2011	.4392	.2308	.3993	.2584	.3656	.2733	.3488
32	.95	.2012	.4597	.2341	.4197	.2571	.3888	.2799	.3629	.2915	.3499
	.99	.1709	.5102	.2113	.4493	.2396	.4094	.2682	.3756	.2828	.3593
33	.95	.2094	.4698	.2443	.4297	.2672	.3998	.2899	.3729	.3015	.3599
	.99	.1792	.5205	.2160	.4594	.2484	.4195	.2775	.3867	.2923	.3693
34	.95	.2177	.4798	.2525	.4397	.2772	.4098	.2989	.3829	.3110	.3699
	.99	.1874	.5308	.2264	.4696	.2586	.4320	.2865	.3967	.3023	.3798
35	.95	.2281	.4899	.2591	.4498	.2855	.4198	.3089	.3929	.3210	.3804
	.99	.1946	.5386	.2367	.4797	.2660	.4421	.2965	.4067	.3118	.3898
36	.95	.2384	.5000	.2692	.4598	.2947	.4298	.3189	.4031	.3305	.3904
	.99	.2017	.5464	.2442	.4898	.2761	.4521	.3056	.4167	.3213	.3998
37	.95	.2460	.5101	.2793	.4698	.3047	.4399	.3279	.4139	.3405	.4004
	.99	.2113	.5551	.2516	.5000	.2862	.4622	.3156	.4268	.3313	.4103
38	.95	.2537	.5201	.2894	.4799	.3148	.4499	.3379	.4239	.3500	.4104
	.99	.2209	.5638	.2619	.5101	.2937	.4723	.3256	.4378	.3408	.4204
39	.95	.2609	.5302	.2995	.4899	.3248	.4599	.3480	.4339	.3600	.4209
	.99	.2266	.5729	.2722	.5202	.3038	.4823	.3346	.4478	.3509	.4304
40	.95	.2680	.5402	.3056	.5000	.3322	.4699	.3576	.4439	.3700	.4309
	.99	.2322	.5820	.2796	.5303	.3139	.4924	.3447	.4578	.3604	.4404
41	.95	.2782	.5503	.3143	.5100	.3423	.4799	.3670	.4539	.3795	.4409
	.99	.2429	.5923	.2871	.5405	.3229	.5025	.3537	.4679	.3704	.4509
42	.95	.2885	.5603	.3244	.5200	.3523	.4899	.3770	.4639	.3895	.4509
	.99	.2536	.6026	.2974	.5506	.3314	.5125	.3637	.4779	.3799	.4609
43	.95	.2986	.5704	.3345	.5301	.3623	.5000	.3870	.4739	.3995	.4609
	.99	.2616	.6129	.3076	.5607	.3415	.5226	.3737	.4879	.3899	.4709
44	.95	.3088	.5805	.3445	.5401	.3724	.5100	.3970	.4839	.4095	.4709
	.99	.2696	.6232	.3178	.5709	.3516	.5327	.3836	.4979	.3999	.4809
45	.95	.3177	.5906	.3546	.5501	.3824	.5200	.4060	.4939	.4190	.4809
	.99	.2767	.6336	.3247	.5810	.3616	.5427	.3928	.5080	.4094	.4909
46	.95	.3266	.6007	.3646	.5602	.3898	.5300	.4160	.5040	.4290	.4909
	.99	.2839	.6440	.3328	.5912	.3692	.5528	.4028	.5180	.4194	.5010
47	.95	.3326	.6108	.3742	.5702	.3999	.5400	.4260	.5140	.4390	.5010
	.99	.2944	.6544	.3430	.6013	.3793	.5629	.4128	.5280	.4295	.5110
48	.95	.3386	.6209	.3814	.5802	.4099	.5500	.4360	.5240	.4490	.5110
	.99	.3049	.6648	.3531	.6103	.3893	.5714	.4229	.5380	.4395	.5210
49	.95	.3488	.6310	.3896	.5903	.4199	.5600	.4460	.5340	.4590	.5210
	.99	.3146	.6703	.3633	.6173	.3994	.5804	.4319	.5481	.4492	.5310
50	.95	.3589	.6411	.3997	.6003	.4299	.5701	.4560	.5440	.4690	.5310
	.99	.3243	.6757	.3735	.6265	.4095	.5905	.4419	.5581	.4590	.5410

TABLE **Q**   Critical values for tests of proportions

This table provides critical values to test binomial proportions 0.5, 0.33333, 0.25, 0.05, and 0.01 for every sample size up to 100, for 110 to 500 in increments of 10, and 550 to 1,000 in increments of 50. The critical values correspond to two-tailed significance levels (α) of 0.1, 0.05, 0.02, and 0.01. For sample sizes less than or equal to 100, the lower (L) and upper (U) critical values are given as integers. The null hypothesis is rejected at the given probability if the observed counts are equal to or more deviant than the tabled values. A dash indicates that no such values exist. For sample sizes larger than 100 the lower and upper critical values are given as percentages to facilitate interpolation.

For example, the critical values to test a proportion of $p = 0.25$ (a 1:3 ratio) for $n = 80$ are 12 and 29 at the 5% level of significance. Samples with as few or fewer than 12, or 29 or more of the specified event would cause one to reject the null hypothesis that $p = 0.25$ (at the 5% level).

The table of critical values for a proportion of 0.5 is used for the sign test (Sections 13.12 and 17.7). The table is also used in Section 17.1 to test for particular proportions (such as those expected in genetics).

This table was computed by summing individual terms of the binomial distribution.

TABLE Q Critical values for tests of proportions

Proportion = 0.5

	0.1		0.05		0.02		0.01	
n	L	U	L	U	L	U	L	U
5	0	5	—	—	—	—	—	—
6	0	6	0	6	—	—	—	—
7	0	7	0	7	0	7	—	—
8	1	7	0	8	0	8	0	8
9	1	8	1	8	0	9	0	9
10	1	9	1	9	0	10	0	10
11	2	9	1	10	1	10	0	11
12	2	10	2	10	1	11	1	11
13	3	10	2	11	1	12	1	12
14	3	11	2	12	2	12	1	13
15	3	12	3	12	2	13	2	13
16	4	12	3	13	2	14	2	14
17	4	13	4	13	3	14	2	15
18	5	13	4	14	3	15	3	15
19	5	14	4	15	4	15	3	16
20	5	15	5	15	4	16	3	17
21	6	15	5	16	4	17	4	17
22	6	16	5	17	5	17	4	18
23	7	16	6	17	5	18	4	19
24	7	17	6	18	5	19	4	19
25	7	18	7	18	6	19	5	19
26	8	18	7	19	6	20	5	20
27	8	19	7	20	7	20	6	20
28	9	19	8	20	7	21	6	21
29	9	20	8	21	7	22	6	22
30	10	20	9	21	8	22	7	22
31	10	21	9	22	8	23	7	23
32	10	22	9	23	8	24	7	24
33	11	22	10	23	9	24	8	24
34	11	23	10	24	9	25	8	25
35	12	23	11	24	10	25	9	25
36	12	24	11	25	10	26	9	26
37	13	24	12	25	10	27	9	27
38	13	25	12	26	11	27	10	27
39	13	26	12	27	11	28	10	28
40	14	26	13	27	12	28	11	28
41	14	27	13	28	12	29	11	29
42	15	27	14	28	13	29	11	30
43	15	28	14	29	13	30	12	30
44	16	28	15	29	13	31	12	31
45	16	29	15	30	14	31	13	31
46	16	30	15	31	14	32	13	32
47	17	30	16	31	15	32	13	33
48	17	31	16	32	15	33	14	33
49	18	31	17	32	15	34	14	34
50	18	32	17	33	16	34	15	34
51	19	32	18	33	16	35	15	35
52	19	33	18	34	17	35	15	36
53	20	33	18	35	17	36	16	36
54	20	34	19	35	18	36	16	37
55	20	35	19	36	18	37	17	37
56	21	35	20	36	18	38	17	38
57	21	36	20	37	19	38	17	39
58	22	36	21	37	19	39	18	39
59	22	37	21	38	20	39	18	40

TABLE Q Critical values for tests of proportions (*continued*)

Proportion = 0.5

α

n	0.1		0.05		0.02		0.01	
	L	**U**	**L**	**U**	**L**	**U**	**L**	**U**
60	23	37	21	39	20	40	19	41
61	23	38	22	39	20	41	20	41
62	24	38	22	40	21	41	20	42
63	24	39	23	40	21	42	20	43
64	24	40	23	41	22	42	21	43
65	25	40	24	41	22	43	21	44
66	25	41	24	42	23	43	22	44
67	26	41	25	42	23	44	22	45
68	26	42	25	43	23	45	22	46
69	27	42	25	44	24	45	23	46
70	27	43	26	44	24	46	23	47
71	28	43	26	45	25	46	24	47
72	28	44	27	45	25	47	24	48
73	28	45	27	46	26	47	25	48
74	29	45	28	46	26	48	25	49
75	29	46	28	47	26	49	25	50
76	30	46	28	48	27	49	26	50
77	30	47	29	48	27	50	26	51
78	31	47	29	49	28	50	27	51
79	31	48	30	49	28	51	27	52
80	32	48	30	50	29	51	28	52
81	32	49	31	50	29	52	28	53
82	33	49	31	51	30	52	28	54
83	33	50	32	51	30	53	29	54
84	33	51	32	52	30	54	29	55
85	34	51	32	53	31	54	30	55
86	34	52	33	53	31	55	30	56
87	35	52	33	54	32	55	31	56
88	35	53	34	54	32	56	31	57
89	36	53	34	55	33	56	31	58
90	36	54	35	55	33	57	32	58
91	37	54	35	56	33	58	32	59
92	37	55	36	56	34	58	33	59
93	38	55	36	57	34	59	33	60
94	38	56	37	57	35	59	34	60
95	38	57	37	58	35	60	34	61
96	39	57	37	59	36	60	34	62
97	39	58	38	59	36	61	35	62
98	40	58	38	60	37	61	35	63
99	40	59	39	60	37	62	36	63
100	41	59	39	61	37	63	36	64

TABLE Q Critical values for tests of proportions (*continued*)

Proportion = 0.5

α

	0.1		0.05		0.02		0.01	
n	L	U	L	U	L	U	L	U
100	41.000	59.000	39.000	61.000	37.000	63.000	36.000	64.000
110	40.909	59.091	40.000	60.000	38.182	61.818	37.273	62.727
120	41.667	58.333	40.000	60.000	38.333	61.667	37.500	62.500
130	42.308	57.692	40.769	59.231	39.231	60.769	37.692	62.308
140	42.143	57.857	40.714	59.286	39.286	60.714	38.571	61.429
150	42.667	57.333	41.333	58.667	40.000	60.000	38.667	61.333
160	43.125	56.875	41.875	58.125	40.000	60.000	39.375	60.625
170	42.941	57.059	41.765	58.235	40.588	59.412	39.412	60.588
180	43.333	56.667	42.222	57.778	40.556	59.444	40.000	60.000
190	43.684	56.316	42.632	57.368	41.053	58.947	40.000	60.000
200	43.500	56.500	42.500	57.500	41.500	58.500	40.500	59.500
210	43.810	56.190	42.857	57.143	41.429	58.571	40.476	59.524
220	44.091	55.909	42.727	57.273	41.818	58.182	40.909	59.091
230	44.348	55.652	43.043	56.957	41.739	58.261	41.304	58.696
240	46.167	55.833	43.333	56.667	42.083	57.917	41.250	58.750
250	44.400	55.600	43.600	56.400	42.400	57.600	41.600	58.400
260	44.615	55.385	43.462	56.538	42.308	57.692	41.538	58.462
270	44.444	55.556	43.704	56.296	42.593	57.407	41.852	58.148
280	44.643	55.357	43.929	56.071	42.857	57.143	41.786	58.214
290	44.828	55.172	43.793	56.207	42.759	57.241	42.069	57.931
300	45.000	55.000	44.000	56.000	43.000	57.000	42.333	57.667
310	45.161	54.839	44.194	55.806	43.226	56.774	42.258	57.742
320	45.000	55.000	44.063	55.938	43.125	56.875	42.500	57.500
330	45.152	54.848	44.242	55.758	43.333	56.667	42.727	57.273
340	45.294	54.706	44.412	55.588	43.529	56.471	42.647	57.353
350	45.429	54.571	44.571	55.429	43.429	56.571	42.857	57.143
360	45.278	54.722	44.444	55.556	43.611	56.389	43.056	56.944
370	45.405	54.595	44.595	55.405	43.784	56.216	42.973	57.027
380	45.526	54.474	44.737	55.263	43.684	56.316	43.158	56.842
390	45.641	54.359	44.872	55.128	43.846	56.154	43.333	56.667
400	45.750	54.250	44.750	55.250	44.000	56.000	43.250	56.750
410	45.610	54.390	44.878	55.122	43.902	56.098	43.415	56.585
420	45.714	54.286	45.000	55.000	44.048	55.952	43.571	56.429
430	45.814	54.186	45.116	54.884	44.186	55.814	43.488	56.512
440	45.909	54.091	45.000	55.000	44.318	55.682	43.636	56.364
450	46.000	54.000	45.111	54.889	44.222	55.778	43.778	56.222
460	45.870	54.130	45.217	54.783	44.348	55.652	43.696	56.304
470	45.957	54.043	45.319	54.681	44.468	55.532	43.830	56.170
480	46.042	53.958	45.417	54.583	44.583	55.417	43.958	56.042
490	46.122	53.878	45.306	54.694	44.490	55.510	44.082	55.918
500	46.200	53.800	45.400	54.600	44.600	55.400	44.000	56.000
550	46.364	53.636	45.636	54.364	44.909	55.091	44.364	55.636
600	46.500	53.500	45.833	54.167	45.167	54.833	44.500	55.500
650	46.615	53.385	46.000	54.000	45.231	54.769	44.769	55.231
700	46.714	53.286	46.143	53.857	45.429	54.571	45.000	55.000
750	46.800	53.200	46.267	53.733	45.600	54.400	45.200	54.800
800	47.000	53.000	46.375	53.625	45.750	54.250	45.375	54.625
850	47.059	52.941	46.471	53.529	45.882	54.118	45.412	54.588
900	47.111	52.889	46.667	53.333	46.000	54.000	45.556	54.444
950	47.263	52.737	46.737	53.263	46.105	53.895	45.684	54.316
1000	47.300	52.700	46.800	53.200	46.200	53.800	45.800	54.200

TABLE Q   Critical values for tests of proportions (*continued*)

Proportion = 0.33333

	α							
	0.1		**0.05**		**0.02**		**0.01**	
n	L	U	L	U	L	U	L	U
3	—	3	—	—	—	—	—	—
4	—	4	—	4	—	4	—	—
5	—	4	—	5	—	5	—	5
6	—	5	—	5	—	6	—	6
7	—	5	—	6	—	6	—	7
8	0	6	—	6	—	7	—	7
9	0	6	—	7	—	7	—	8
10	0	7	0	7	—	8	—	8
11	0	7	0	8	—	8	—	9
12	0	8	0	8	0	9	—	9
13	1	8	0	9	0	9	—	10
14	1	9	0	9	0	10	0	10
15	1	9	1	10	0	10	0	11
16	1	9	1	10	0	11	0	11
17	2	10	1	11	1	11	0	12
18	2	10	1	11	1	12	0	12
19	2	11	2	11	1	12	1	13
20	2	11	2	12	1	13	1	13
21	3	12	2	12	1	13	1	14
22	3	12	2	13	2	14	1	14
23	3	12	2	13	2	14	1	15
24	3	13	3	14	2	15	2	15
25	4	13	3	14	2	15	2	16
26	4	14	3	14	2	15	2	16
27	4	14	3	15	3	16	2	17
28	4	15	4	15	3	16	2	17
29	5	15	4	16	3	17	3	17
30	5	15	4	16	3	17	3	18
31	5	16	4	17	4	18	3	18
32	5	16	5	17	4	18	3	19
33	6	17	5	17	4	18	3	19
34	6	17	5	18	4	19	4	20
35	6	17	5	18	5	19	4	20
36	6	18	6	19	5	20	4	20
37	7	18	6	19	5	20	4	21
38	7	19	6	19	5	21	5	21
39	7	19	6	20	5	21	5	22
40	8	19	7	20	6	21	5	22
41	8	20	7	21	6	22	5	23
42	8	20	7	21	6	22	6	23
43	8	20	7	22	6	23	6	24
44	9	21	8	22	7	23	6	24
45	9	21	8	22	7	24	6	24
46	9	22	8	23	7	24	7	25
47	9	22	9	23	7	24	7	25
48	10	22	9	24	8	25	7	26
49	10	23	9	24	8	25	7	26
50	10	23	9	24	8	26	7	26
51	11	24	10	25	8	26	8	27
52	11	24	10	25	9	26	8	27
53	11	24	10	26	9	27	8	28
54	11	25	10	26	9	27	8	28
55	12	25	11	26	10	28	9	29

TABLE Q   Critical values for tests of proportions (*continued*)

Proportion = 0.33333

α

n	0.1 L	0.1 U	0.05 L	0.05 U	0.02 L	0.02 U	0.01 L	0.01 U
56	12	26	11	27	10	28	9	29
57	12	26	11	27	10	28	9	29
58	13	26	11	27	10	29	9	30
59	13	27	12	28	11	29	10	30
60	13	27	12	28	11	30	10	31
61	13	27	12	29	11	30	10	31
62	14	28	13	29	11	30	11	31
63	14	28	13	29	12	31	11	32
64	14	29	13	30	12	31	11	32
65	15	29	13	30	12	32	11	33
66	15	29	14	31	12	32	12	33
67	15	30	14	31	13	33	12	34
68	15	30	14	31	13	33	12	34
69	16	31	15	32	13	33	12	34
70	16	31	15	32	13	34	13	35
71	16	31	15	33	14	34	13	35
72	17	32	15	33	14	34	13	36
73	17	32	16	33	14	35	13	36
74	17	32	16	34	15	35	14	36
75	17	33	16	34	15	36	14	37
76	18	33	16	35	15	36	14	37
77	18	34	17	35	15	36	14	38
78	18	34	17	35	16	37	15	38
79	19	34	17	36	16	37	15	38
80	19	35	18	36	16	38	15	39
81	19	35	18	36	16	38	15	39
82	19	35	18	37	17	38	16	40
83	20	36	18	37	17	39	16	40
84	20	36	19	38	17	39	16	40
85	20	37	19	38	18	40	17	41
86	21	37	19	38	18	40	17	41
87	21	37	20	39	18	40	17	42
88	21	38	20	39	18	41	17	42
89	21	38	20	40	19	41	18	42
90	22	38	20	40	19	42	18	43
91	22	39	21	40	19	42	18	43
92	22	39	21	41	19	42	18	44
93	23	40	21	41	20	43	19	44
94	23	40	22	41	20	43	19	44
95	23	40	22	42	20	44	19	45
96	24	41	22	42	21	44	19	45
97	24	41	22	43	21	44	20	46
98	24	41	23	43	21	45	20	46
99	24	42	23	43	21	45	20	46
100	25	42	23	44	22	46	21	47

TABLE Q   Critical values for tests of proportions (*continued*)

Proportion = 0.33333

α

n	0.1 L	0.1 U	0.05 L	0.05 U	0.02 L	0.02 U	0.01 L	0.01 U
100	25.000	42.000	23.000	44.000	22.000	46.000	21.000	47.000
110	25.455	41.818	23.636	42.727	21.818	44.545	20.909	46.364
120	25.833	41.667	24.167	42.500	22.500	44.167	21.667	45.833
130	26.154	40.769	24.615	42.308	23.077	43.846	22.308	44.615
140	26.429	40.714	25.000	42.143	23.571	43.571	22.857	44.286
150	26.667	40.667	25.333	41.333	24.000	43.333	22.667	44.000
160	26.875	40.000	25.625	41.250	24.375	42.500	23.125	43.750
170	27.059	40.000	25.882	41.176	24.706	42.353	23.529	43.529
180	27.222	39.444	26.111	41.111	25.000	42.222	23.889	43.333
190	27.368	39.474	26.316	40.526	24.737	42.105	24.211	42.632
200	27.500	39.500	26.500	40.500	25.000	41.500	24.500	42.500
210	27.619	39.048	26.667	40.476	25.238	41.429	24.762	42.381
220	27.727	39.091	26.818	40.000	25.455	41.364	25.000	42.273
230	27.826	39.130	26.957	40.000	25.652	41.304	25.217	41.739
240	27.917	38.750	27.083	39.583	25.833	40.833	25.417	41.667
250	28.000	38.800	27.200	39.600	26.000	40.800	25.200	41.600
260	28.077	38.462	27.308	39.615	26.154	40.769	25.385	41.538
270	28.148	38.519	27.407	39.259	26.296	40.370	25.556	41.111
280	28.214	38.214	27.500	39.286	26.429	40.357	25.714	41.071
290	28.621	38.276	27.586	39.310	26.552	40.345	25.862	41.034
300	28.667	38.333	27.667	39.000	26.667	40.000	26.000	40.667
310	28.710	38.065	27.742	39.032	26.774	40.000	26.129	40.645
320	28.750	38.125	27.813	38.750	26.875	40.000	26.250	40.625
330	28.788	37.879	27.879	38.788	26.970	39.697	26.364	40.303
340	28.824	37.941	27.941	38.824	27.059	39.706	26.471	40.294
350	28.857	37.714	28.286	38.571	27.143	39.429	26.571	40.286
360	28.889	37.778	28.333	38.611	27.222	39.444	26.667	40.000
370	29.189	37.568	28.378	38.378	27.568	39.459	26.757	40.000
380	29.211	37.632	28.421	38.421	27.632	39.211	26.842	40.000
390	29.231	37.436	28.462	38.205	27.692	39.231	26.923	39.744
400	29.250	37.500	28.500	38.250	27.750	39.000	27.000	39.750
410	29.268	37.317	28.537	38.293	27.805	39.024	27.073	39.756
420	29.286	37.381	28.571	38.095	27.857	39.048	27.143	39.524
430	29.302	37.442	28.605	38.140	27.907	38.837	27.209	39.535
440	29.545	37.273	28.636	37.955	27.955	38.864	27.500	39.318
450	29.556	37.333	28.889	38.000	28.000	38.667	27.556	39.333
460	29.565	37.174	28.913	37.826	28.043	38.696	27.609	39.348
470	29.574	37.234	28.936	37.872	28.085	38.723	27.660	39.149
480	29.583	37.083	28.958	37.708	28.125	38.542	27.708	39.167
490	29.592	37.143	28.980	37.755	28.163	38.571	27.755	38.980
500	29.600	37.000	29.000	37.600	28.200	38.400	27.800	39.000
550	29.818	36.909	29.273	37.455	28.545	38.182	28.000	38.727
600	30.000	36.667	29.500	37.333	28.667	38.000	28.333	38.500
650	30.154	36.615	29.538	37.077	28.923	37.846	28.462	38.308
700	30.286	36.429	29.714	37.000	29.143	37.714	28.714	38.143
750	30.400	36.267	29.867	36.800	29.200	37.467	28.800	38.000
800	30.500	36.250	30.000	36.750	29.375	37.375	29.000	37.750
850	30.588	36.118	30.118	36.588	29.529	37.294	29.059	37.647
900	30.667	36.000	30.111	36.556	29.556	37.111	29.222	37.556
950	30.737	36.000	30.211	36.421	29.684	37.053	29.368	37.368
1000	30.800	35.900	30.300	36.400	29.800	36,900	29.400	37.300

TABLE Q   Critical values for tests of proportions (*continued*)

Proportion = 0.25

	α							
	0.1		**0.05**		**0.02**		**0.01**	
n	L	U	L	U	L	U	L	U
3	—	3	—	3	—	—	—	—
4	—	4	—	4	—	4	—	4
5	—	4	—	4	—	5	—	5
6	—	4	—	5	—	5	—	5
7	—	5	—	5	—	6	—	6
8	—	5	—	6	—	6	—	6
9	—	5	—	6	—	6	—	7
10	—	6	—	6	—	7	—	7
11	0	6	—	7	—	7	—	8
12	0	7	—	7	—	8	—	8
13	0	7	0	7	—	8	—	9
14	0	7	0	8	—	9	—	9
15	0	8	0	8	—	9	—	9
16	0	8	0	9	—	9	—	10
17	0	8	0	9	0	10	—	10
18	1	9	0	9	0	10	—	11
19	1	9	0	10	0	10	0	11
20	1	9	1	10	0	11	0	11
21	1	10	1	10	0	11	0	12
22	1	10	1	11	0	11	0	12
23	2	10	1	11	0	12	0	12
24	2	11	1	11	1	12	0	13
25	2	11	1	12	1	13	0	13
26	2	11	1	12	1	13	0	14
27	2	12	2	12	1	13	1	14
28	2	12	2	13	1	14	1	14
29	3	12	2	13	1	14	1	15
30	3	13	2	13	1	14	1	15
31	3	13	2	14	2	15	1	15
32	3	13	2	14	2	15	1	16
33	3	13	3	14	2	15	1	16
34	4	14	3	15	2	16	2	16
35	4	14	3	15	2	16	2	17
36	4	14	3	15	2	16	2	17
37	4	15	3	16	3	17	2	17
38	4	15	4	16	3	17	2	18
39	4	15	4	16	3	17	2	18
40	5	16	4	17	3	18	3	18
41	5	16	4	17	3	18	3	19
42	5	16	4	17	3	18	3	19
43	5	17	4	18	4	19	3	19
44	5	17	5	18	4	19	3	20
45	6	17	5	18	4	19	3	20
46	6	17	5	18	4	20	4	20
47	6	18	5	19	4	20	4	21
48	6	18	5	19	4	20	4	21
49	6	18	6	19	5	21	4	21
50	7	19	6	20	5	21	4	22
51	7	19	6	20	5	21	4	22
52	7	19	6	20	5	22	5	22
53	7	20	6	21	5	22	5	23
54	7	20	7	21	6	22	5	23
55	8	20	7	21	6	23	5	23

TABLE Q Critical values for tests of proportions (*continued*)

Proportion = 0.25

α

	0.1		0.05		0.02		0.01	
n	L	U	L	U	L	U	L	U
56	8	20	7	22	6	23	5	24
57	8	21	7	22	6	23	5	24
58	8	21	7	22	6	23	6	24
59	8	21	8	22	6	24	6	25
60	9	22	8	23	7	24	6	25
61	9	22	8	23	7	24	6	25
62	9	22	8	23	7	25	6	26
63	9	23	8	24	7	25	6	26
64	9	23	8	24	7	25	7	26
65	10	23	9	24	8	26	7	27
66	10	23	9	25	8	26	7	27
67	10	24	9	25	8	26	7	27
68	10	24	9	25	8	27	7	28
69	10	24	9	26	8	27	8	28
70	11	25	10	26	9	27	8	28
71	11	25	10	26	9	28	8	29
72	11	25	10	26	9	28	8	29
73	11	25	10	27	9	28	8	29
74	12	26	10	27	9	28	8	29
75	12	26	11	27	9	29	9	30
76	12	26	11	28	10	29	9	30
77	12	27	11	28	10	29	9	30
78	12	27	11	28	10	30	9	31
79	13	27	11	28	10	30	9	31
80	13	27	12	29	10	30	10	31
81	13	28	12	29	11	31	10	32
82	13	28	12	29	11	31	10	32
83	13	28	12	30	11	31	10	32
84	14	29	12	30	11	32	10	33
85	14	29	13	30	11	32	11	33
86	14	29	13	31	12	32	11	33
87	14	30	13	31	12	32	11	34
88	14	30	13	31	12	33	11	34
89	15	30	14	31	12	33	11	34
90	15	30	14	32	12	33	11	34
91	15	31	14	32	13	34	12	35
92	15	31	14	32	13	34	12	35
93	16	31	14	33	13	34	12	35
94	16	32	15	33	13	35	12	36
95	16	32	15	33	13	35	12	36
96	16	32	15	34	14	35	13	36
97	16	32	15	34	14	35	13	37
98	17	33	15	34	14	36	13	37
99	17	33	16	34	14	36	13	37
100	17	33	16	35	14	36	13	38

TABLE **Q**   Critical values for tests of proportions (*continued*)

Proportion = 0.25

	α							
	0.1		**0.05**		**0.02**		**0.01**	
n	L	U	L	U	L	U	L	U
100	17.000	33.000	16.000	35.000	14.000	36.000	13.000	38.000
110	17.273	32.727	16.364	34.545	14.545	35.455	13.636	37.273
120	17.500	32.500	16.667	34.167	15.000	35.000	14.167	36.667
130	18.462	32.308	16.923	33.077	15.385	34.615	14.615	36.154
140	18.571	32.143	17.143	32.857	15.714	34.286	15.000	35.714
150	18.667	31.333	17.333	32.667	16.667	34.000	15.333	35.333
160	18.750	31.250	18.125	32.500	16.875	33.750	15.625	35.000
170	18.824	31.176	18.235	32.353	17.059	33.529	15.882	34.118
180	19.444	31.111	18.333	32.222	17.222	33.333	16.667	33.889
190	19.474	30.526	18.421	31.579	17.368	33.158	16.842	33.684
200	19.500	30.500	18.500	31.500	17.500	33.000	17.000	33.500
210	19.524	30.476	18.571	31.429	17.619	32.381	17.143	33.333
220	20.000	30.455	19.091	31.364	17.727	32.273	17.273	33.182
230	20.000	30.000	19.130	31.304	18.261	32.174	17.391	33.043
240	20.000	30.000	19.167	30.833	18.333	32.083	17.500	32.917
250	20.000	30.000	19.200	30.800	18.400	32.000	17.600	32.800
260	20.385	30.000	19.615	30.769	18.462	31.923	18.077	32.308
270	20.370	29.630	19.630	30.741	18.519	31.481	18.148	32.222
280	20.357	29.643	19.643	30.357	18.929	31.429	18.214	32.143
290	20.690	29.655	19.655	30.345	18.966	31.379	18.276	32.069
300	20.667	29.333	20.000	30.333	19.000	31.333	18.333	32.000
310	20.645	29.355	20.000	30.323	19.032	31.290	18.387	31.935
320	20.625	29.375	20.000	30.000	19.063	30.938	18.750	31.563
330	20.909	29.394	20.000	30.000	19.394	30.909	18.788	31.515
340	20.882	29.118	20.294	30.000	19.412	30.882	18.824	31.471
350	20.857	29.143	20.286	30.000	19.429	30.857	18.857	31.429
360	21.111	29.167	20.278	29.722	19.444	30.556	18.889	31.389
370	21.081	28.919	20.270	29.730	19.730	30.541	19.189	31.081
380	21.053	28.947	20.526	29.737	19.737	30.526	19.211	31.053
390	21.282	28.974	20.513	29.487	19.744	30.513	19.231	31.026
400	21.250	28.750	20.500	29.500	19.750	30.250	19.250	31.000
410	21.220	28.780	20.732	29.512	19.756	30.244	19.268	30.976
420	21.429	28.810	20.714	29.524	20.000	30.238	19.524	30.714
430	21.395	28.605	20.698	29.302	20.000	30.233	19.535	30.698
440	21.364	28.636	20.682	29.318	20.000	30.000	19.545	30.682
450	21.556	28.667	20.889	29.333	20.222	30.000	19.556	30.667
460	21.522	28.478	20.870	29.130	20.217	30.000	19.783	30.435
470	21.489	28.511	20.851	29.149	20.213	30.000	19.787	30.426
480	21.667	28.542	21.042	29.167	20.208	29.792	19.792	30.417
490	21.633	28.367	21.020	29.184	20.408	29.796	19.796	30.408
500	21.600	28.400	21.000	29.000	20.400	29.800	20.000	30.200
550	21.818	28.182	21.273	28.909	20.545	29.455	20.182	30.000
600	22.000	28.167	21.333	28.667	20.833	29.333	20.333	29.833
650	22.000	28.000	21.538	28.462	20.923	29.231	20.615	29.538
700	22.143	27.857	21.714	28.429	21.143	29.000	20.714	29.429
750	22.267	27.733	21.867	28.267	21.200	28.800	20.800	29.200
800	22.375	27.625	21.875	28.125	21.375	28.750	21.000	29.125
850	22.471	27.529	22.000	28.000	21.529	28.588	21.059	28.941
900	22.556	27.556	22.111	28.000	21.556	28.556	21.222	28.889
950	22.632	27.474	22.211	27.895	21.684	28.421	21.368	28.737
1000	22.700	27.400	22.200	27.800	21.800	28.300	21.400	28.700

TABLE Q Critical values for tests of proportions (*continued*)

Proportion = 0.05

	α							
	0.1		**0.05**		**0.02**		**0.01**	
n	**L**	**U**	**L**	**U**	**L**	**U**	**L**	**U**
2	—	3	—	3	—	—	—	—
3	—	2	—	2	—	2	—	3
4	—	2	—	2	—	3	—	3
5	—	2	—	2	—	3	—	3
6	—	2	—	3	—	3	—	3
7	—	2	—	3	—	3	—	3
8	—	3	—	3	—	3	—	4
9	—	3	—	3	—	3	—	4
10	—	3	—	3	—	4	—	4
11	—	3	—	3	—	4	—	4
12	—	3	—	3	—	4	—	4
13	—	3	—	3	—	4	—	4
14	—	3	—	4	—	4	—	4
15	—	3	—	4	—	4	—	5
16	—	3	—	4	—	4	—	5
17	—	4	—	4	—	4	—	5
18	—	4	—	4	—	5	—	5
19	—	4	—	4	—	5	—	5
20	—	4	—	4	—	5	—	5
21	—	4	—	4	—	5	—	5
22	—	4	—	4	—	5	—	5
23	—	4	—	5	—	5	—	5
24	—	4	—	5	—	5	—	6
25	—	4	—	5	—	5	—	6
26	—	4	—	5	—	5	—	6
27	—	4	—	5	—	6	—	6
28	—	4	—	5	—	6	—	6
29	—	5	—	5	—	6	—	6
30	—	5	—	5	—	6	—	6
31	—	5	—	5	—	6	—	6
32	—	5	—	5	—	6	—	6
33	—	5	—	5	—	6	—	7
34	—	5	—	6	—	6	—	7
35	—	5	—	6	—	6	—	7
36	—	5	—	6	—	6	—	7
37	—	5	—	6	—	6	—	7
38	—	5	—	6	—	7	—	7
39	—	5	—	6	—	7	—	7
40	—	5	—	6	—	7	—	7
41	—	6	—	6	—	7	—	7
42	—	6	—	6	—	7	—	7
43	—	6	—	6	—	7	—	8
44	—	6	—	6	—	7	—	8
45	—	6	—	6	—	7	—	8
46	—	6	—	7	—	7	—	8
47	—	6	—	7	—	7	—	8
48	—	6	—	7	—	7	—	8
49	—	6	—	7	—	8	—	8
50	—	6	—	7	—	8	—	8
51	—	6	—	7	—	8	—	8
52	—	6	—	7	—	8	—	8
53	—	6	—	7	—	8	—	8
54	—	7	—	7	—	8	—	9

TABLE Q   Critical values for tests of proportions (*continued*)

Proportion = 0.05

					α			
	0.1		**0.05**		**0.02**		**0.01**	
n	L	U	L	U	L	U	L	U
55	—	7	—	7	—	8	—	9
56	—	7	—	7	—	8	—	9
57	—	7	—	7	—	8	—	9
58	—	7	—	8	—	8	—	9
59	0	7	—	8	—	8	—	9
60	0	7	—	8	—	8	—	9
61	0	7	—	8	—	9	—	9
62	0	7	—	8	—	9	—	9
63	0	7	—	8	—	9	—	9
64	0	7	—	8	—	9	—	9
65	0	7	—	8	—	9	—	9
66	0	7	—	8	—	9	—	10
67	0	7	—	8	—	9	—	10
68	0	8	—	8	—	9	—	10
69	0	8	—	8	—	9	—	10
70	0	8	—	8	—	9	—	10
71	0	8	—	9	—	9	—	10
72	0	8	0	9	—	9	—	10
73	0	8	0	9	—	10	—	10
74	0	8	0	9	—	10	—	10
75	0	8	0	9	—	10	—	10
76	0	8	0	9	—	10	—	10
77	0	8	0	9	—	10	—	10
78	0	8	0	9	—	10	—	11
79	0	8	0	9	—	10	—	11
80	0	8	0	9	—	10	—	11
81	0	8	0	9	—	10	—	11
82	0	9	0	9	—	10	—	11
83	0	9	0	9	—	10	—	11
84	0	9	0	9	—	10	—	11
85	0	9	0	10	—	10	—	11
86	0	9	0	10	—	11	—	11
87	0	9	0	10	—	11	—	11
88	0	9	0	10	—	11	—	11
89	0	9	0	10	—	11	—	11
90	0	9	0	10	0	11	—	12
91	0	9	0	10	0	11	—	12
92	0	9	0	10	0	11	—	12
93	1	9	0	10	0	11	—	12
94	1	9	0	10	0	11	—	12
95	1	9	0	10	0	11	—	12
96	1	10	0	10	0	11	—	12
97	1	10	0	10	0	11	—	12
98	1	10	0	10	0	11	—	12
99	1	10	0	11	0	12	—	12
100	1	10	0	11	0	12	—	12

TABLE Q Critical values for tests of proportions (*continued*)

Proportion = 0.05

α

	0.1		0.05		0.02		0.01	
n	L	U	L	U	L	U	L	U
100	1.000	10.000	0.000	11.000	0.000	12.000	—	12.000
110	0.909	9.091	0.909	10.000	0.000	10.909	0.000	11.818
120	0.833	9.167	0.833	10.000	0.000	10.833	0.000	11.667
130	1.538	9.231	0.769	10.000	0.769	10.769	0.000	11.538
140	1.429	8.571	0.714	9.286	0.714	10.714	0.000	10.714
150	1.333	8.667	1.333	9.333	0.667	10.000	0.667	10.667
160	1.875	8.750	1.250	9.375	0.625	10.000	0.625	10.625
170	1.765	8.235	1.176	8.824	1.176	10.000	0.588	10.588
180	1.667	8.333	1.667	8.889	1.111	9.444	0.556	10.000
190	2.105	8.421	1.579	8.947	1.053	9.474	1.053	10.000
200	2.000	8.000	1.500	8.500	1.500	9.500	1.000	10.000
210	2.381	8.095	1.905	8.571	1.429	9.048	0.952	9.524
220	2.273	8.182	1.818	8.636	1.364	9.091	1.364	9.545
230	2.174	7.826	2.174	8.261	1.739	9.130	1.304	9.565
240	2.500	7.917	2.083	8.333	1.667	8.750	1.250	9.167
250	2.400	7.600	2.000	8.400	1.600	8.800	1.600	9.200
260	2.692	7.692	2.308	8.077	1.923	8.846	1.538	9.231
270	2.593	7.778	2.222	8.148	1.852	8.519	1.481	8.889
280	2.500	7.500	2.143	8.214	1.786	8.571	1.786	8.929
290	2.759	7.586	2.414	7.931	2.069	8.621	1.724	8.966
300	2.667	7.333	2.333	8.000	2.000	8.333	1.667	8.667
310	2.581	7.419	2.258	7.742	1.935	8.387	1.935	8.710
320	2.813	7.500	2.500	7.813	2.188	8.438	1.875	8.750
330	2.727	7.273	2.424	7.879	2.121	8.182	1.818	8.485
340	2.941	7.353	2.647	7.647	2.059	8.235	2.059	8.529
350	2.857	7.143	2.571	7.714	2.286	8.286	2.000	8.571
360	2.778	7.222	2.500	7.500	2.222	8.056	1.944	8.333
370	2.973	7.297	2.703	7.568	2.162	8.108	2.162	8.378
380	2.895	7.105	2.632	7.632	2.368	7.895	2.105	8.421
390	3.077	7.179	2.821	7.436	2.308	7.949	2.051	8.205
400	3.000	7.000	2.750	7.500	2.500	8.000	2.250	8.250
410	3.171	7.073	2.683	7.561	2.439	7.805	2.195	8.293
420	3.095	7.143	2.857	7.381	2.381	7.857	2.143	8.095
430	3.023	6.977	2.791	7.442	2.558	7.907	2.326	8.140
440	3.182	7.045	2.955	7.273	2.500	7.727	2.273	8.182
450	3.111	6.889	2.889	7.333	2.444	7.778	2.444	8.000
460	3.261	6.957	2.826	7.391	2.609	7.609	2.391	8.043
470	3.191	7.021	2.979	7.234	2.553	7.660	2.340	7.872
480	3.125	6.875	2.917	7.292	2.708	7.708	2.500	7.917
490	3.265	6.939	3.061	7.143	2.653	7.551	2.449	7.959
500	3.200	6.800	3.000	7.200	2.600	7.600	2.400	7.800
550	3.273	6.727	3.091	7.091	2.727	7.455	2.545	7.636
600	3.333	6.667	3.167	7.000	2.833	7.333	2.667	7.667
650	3.538	6.615	3.231	6.923	2.923	7.231	2.769	7.538
700	3.571	6.571	3.286	6.857	3.000	7.143	2.857	7.429
750	3.600	6.533	3.333	6.800	3.067	7.067	2.933	7.333
800	3.625	6.375	3.375	6.625	3.125	7.000	3.000	7.250
850	3.647	6.353	3.412	6.588	3.176	6.941	3.059	7.176
900	3.778	6.333	3.556	6.556	3.222	6.889	3.111	7.111
950	3.789	6.316	3.579	6.526	3.368	6.842	3.158	7.053
1000	3.800	6.300	3.600	6.500	3.400	6.800	3.200	7.000

TABLE Q   Critical values for tests of proportions (*continued*)

Proportion = 0.01

				α				
	0.1		**0.05**		**0.02**		**0.01**	
n	**L**	**U**	**L**	**U**	**L**	**U**	**L**	**U**
5	—	1	—	2	—	2	—	2
6	—	2	—	2	—	2	—	2
7	—	2	—	2	—	2	—	2
8	—	2	—	2	—	2	—	2
9	—	2	—	2	—	2	—	2
10	—	2	—	2	—	2	—	2
11	—	2	—	2	—	2	—	3
12	—	2	—	2	—	2	—	3
13	—	2	—	2	—	2	—	3
14	—	2	—	2	—	2	—	3
15	—	2	—	2	—	2	—	3
16	—	2	—	2	—	3	—	3
17	—	2	—	2	—	3	—	3
18	—	2	—	2	—	3	—	3
19	—	2	—	2	—	3	—	3
20	—	2	—	2	—	3	—	3
21	—	2	—	2	—	3	—	3
22	—	2	—	2	—	3	—	3
23	—	2	—	2	—	3	—	3
24	—	2	—	2	—	3	—	3
25	—	2	—	3	—	3	—	3
26	—	2	—	3	—	3	—	3
27	—	2	—	3	—	3	—	3
28	—	2	—	3	—	3	—	3
29	—	2	—	3	—	3	—	3
30	—	2	—	3	—	3	—	3
31	—	2	—	3	—	3	—	3
32	—	2	—	3	—	3	—	3
33	—	2	—	3	—	3	—	3
34	—	2	—	3	—	3	—	3
35	—	2	—	3	—	3	—	4
36	—	3	—	3	—	3	—	4
37	—	3	—	3	—	3	—	4
38	—	3	—	3	—	3	—	4
39	—	3	—	3	—	3	—	4
40	—	3	—	3	—	3	—	4
41	—	3	—	3	—	3	—	4
42	—	3	—	3	—	3	—	4
43	—	3	—	3	—	3	—	4
44	—	3	—	3	—	3	—	4
45	—	3	—	3	—	4	—	4
46	—	3	—	3	—	4	—	4
47	—	3	—	3	—	4	—	4
48	—	3	—	3	—	4	—	4
49	—	3	—	3	—	4	—	4
50	—	3	—	3	—	4	—	4
51	—	3	—	3	—	4	—	4
52	—	3	—	3	—	4	—	4
53	—	3	—	3	—	4	—	4
54	—	3	—	3	—	4	—	4
55	—	3	—	3	—	4	—	4
56	—	3	—	3	—	4	—	4

TABLE Q Critical values for tests of proportions (*continued*)

Proportion = 0.05

α

	0.1		0.05		0.02		0.01	
n	L	U	L	U	L	U	L	U
57	—	3	—	3	—	4	—	4
58	—	3	—	3	—	4	—	4
59	—	3	—	3	—	4	—	4
60	—	3	—	3	—	4	—	4
61	—	3	—	3	—	4	—	4
62	—	3	—	3	—	4	—	4
63	—	3	—	4	—	4	—	4
64	—	3	—	4	—	4	—	4
65	—	3	—	4	—	4	—	4
66	—	3	—	4	—	4	—	4
67	—	3	—	4	—	4	—	4
68	—	3	—	4	—	4	—	4
69	—	3	—	4	—	4	—	5
70	—	3	—	4	—	4	—	5
71	—	3	—	4	—	4	—	5
72	—	3	—	4	—	4	—	5
73	—	3	—	4	—	4	—	5
74	—	3	—	4	—	4	—	5
75	—	3	—	4	—	4	—	5
76	—	3	—	4	—	4	—	5
77	—	3	—	4	—	4	—	5
78	—	3	—	4	—	4	—	5
79	—	3	—	4	—	4	—	5
80	—	3	—	4	—	4	—	5
81	—	3	—	4	—	4	—	5
82	—	3	—	4	—	4	—	5
83	—	4	—	4	—	4	—	5
84	—	4	—	4	—	5	—	5
85	—	4	—	4	—	5	—	5
86	—	4	—	4	—	5	—	5
87	—	4	—	4	—	5	—	5
88	—	4	—	4	—	5	—	5
89	—	4	—	4	—	5	—	5
90	—	4	—	4	—	5	—	5
91	—	4	—	4	—	5	—	5
92	—	4	—	4	—	5	—	5
93	—	4	—	4	—	5	—	5
94	—	4	—	4	—	5	—	5
95	—	4	—	4	—	5	—	5
96	—	4	—	4	—	5	—	5
97	—	4	—	4	—	5	—	5
98	—	4	—	4	—	5	—	5
99	—	4	—	4	—	5	—	5
100	—	4	—	4	—	5	—	5

TABLE Q Critical values for tests of proportions (*continued*)

Proportion = 0.01

α

	0.1		0.05		0.02		0.01	
n	L	U	L	U	L	U	L	U
100	—	4.000	—	4.000	—	5.000	—	5.000
110	—	3.636	—	4.545	—	4.545	—	5.455
120	—	3.333	—	4.167	—	4.167	—	5.000
130	—	3.077	—	3.846	—	4.615	—	4.615
140	—	3.571	—	3.571	—	4.286	—	4.286
150	—	3.333	—	3.333	—	4.000	—	4.000
160	—	3.125	—	3.125	—	3.750	—	4.375
170	—	2.941	—	3.529	—	3.529	—	4.118
180	—	2.778	—	3.333	—	3.333	—	3.889
190	—	2.632	—	3.158	—	3.684	—	3.684
200	—	3.000	—	3.000	—	3.500	—	3.500
210	—	2.857	—	2.857	—	3.333	—	3.810
220	—	2.727	—	2.727	—	3.182	—	3.636
230	—	2.609	—	3.043	—	3.043	—	3.478
240	—	2.500	—	2.917	—	3.333	—	3.333
250	—	2.400	—	2.800	—	3.200	—	3.200
260	—	2.308	—	2.692	—	3.077	—	3.462
270	—	2.593	—	2.593	—	2.963	—	3.333
280	—	2.500	—	2.500	—	2.857	—	3.214
290	—	2.414	—	2.759	—	2.759	—	3.103
300	0.000	2.333	—	2.667	—	3.000	—	3.000
310	0.000	2.258	—	2.581	—	2.903	—	2.903
320	0.000	2.188	—	2.500	—	2.813	—	3.125
330	0.000	2.424	—	2.424	—	2.727	—	3.030
340	0.000	2.353	—	2.353	—	2.647	—	2.941
350	0.000	2.286	—	2.571	—	2.571	—	2.857
360	0.000	2.222	—	2.500	—	2.778	—	2.778
370	0.000	2.162	0.000	2.432	—	2.703	—	2.703
380	0.000	2.105	0.000	2.368	—	2.632	—	2.895
390	0.000	2.051	0.000	2.308	—	2.564	—	2.821
400	0.000	2.250	0.000	2.250	—	2.500	—	2.750
410	0.000	2.195	0.000	2.195	—	2.439	—	2.683
420	0.000	2.143	0.000	2.381	—	2.619	—	2.619
430	0.000	2.093	0.000	2.326	—	2.558	—	2.558
440	0.000	2.045	0.000	2.273	—	2.500	—	2.727
450	0.000	2.000	0.000	2.222	—	2.444	—	2.667
460	0.000	1.957	0.000	2.174	0.000	2.391	—	2.609
470	0.000	1.915	0.000	2.128	0.000	2.340	—	2.553
480	0.208	2.083	0.000	2.083	0.000	2.500	—	2.500
490	0.204	2.041	0.000	2.245	0.000	2.449	—	2.449
500	0.200	2.000	0.000	2.200	0.000	2.400	—	2.600
550	0.182	2.000	0.000	2.000	0.000	2.364	0.000	2.364
600	0.167	1.833	0.167	2.000	0.000	2.167	0.000	2.333
650	0.308	1.846	0.154	2.000	0.000	2.154	0.000	2.308
700	0.286	1.857	0.143	2.000	0.143	2.143	0.000	2.286
750	0.267	1.733	0.267	1.867	0.133	2.000	0.133	2.133
800	0.375	1.750	0.250	1.875	0.125	2.000	0.125	2.125
850	0.353	1.765	0.235	1.882	0.235	2.000	0.118	2.118
900	0.333	1.667	0.333	1.778	0.222	2.000	0.111	2.111
950	0.421	1.684	0.316	1.789	0.211	1.895	0.211	2.000
1000	0.400	1.600	0.300	1.800	0.200	1.900	0.200	2.000

TABLE **R** Critical values for correlation coefficients

This table furnishes 0.05 and 0.01 critical values for product–moment correlation coefficients r and multiple correlation coefficients R. These values are given for every degree of freedom between $v = 1$ and $v = 30$ and selected degrees of freedom between $v = 30$ and $v = 1000$. This table is used to test the null hypothesis that the correlation coefficient of the population from which the sample has been taken is zero.

Under such conditions the following t-test for significance of the product–moment correlation coefficient r applies:

$$t = r/\sqrt{(1 - r^2)/(n - 2)} = r\sqrt{(n - 2)/(1 - r^2)} \quad v = n - 2$$

where n is the sample size (number of pairs of variates). The critical values in the table are computed by entering the correct values of $t_{\alpha[v]}$ and n in this equation and solving for r.

The critical values for the multiple correlation coefficients are based on a different formula involving the F-distribution:

$$R_{Y \cdot 1 \ldots k(\alpha)} = \left[\frac{k\, F_{\alpha[k, \, n - m]}}{n - m + k\, F_{\alpha[k, \, n - m]}} \right]^{1/2}$$

where $R_{Y \cdot 1 \ldots k(\alpha)}$ is the multiple correlation coefficient of Y with k other variables, n is the sample size, m equals $k + 1$, $F_{\alpha[k, \, n - m]}$ is the critical value of the F-distribution as shown in Table **F**, and α is the desired probability.

To test the significance of a correlation coefficient, the sample size n upon which it is based must be known. Enter the table at $v = n - 2$ degrees of freedom and consult the first column of values headed "number of independent variables." For example, for a sample size of $n = 28$ and $v = 28 - 2 = 26$, the critical values of r are found to be 0.374 at the 5% level and 0.478 at the 1% level. Thus, for an observed correlation coefficient $r = 0.31$ in a sample of 28 paired observations, one would be led to conclude that the correlation between the variables concerned is not significantly different from zero. Negative correlations are considered as positive for purposes of this test. More details on significance testing of correlation coefficients are given in Section 15.5 and Box 15.4.

The other three columns in the table give critical values for a multiple correlation involving 2, 3, and 4 independent variables. Degrees of freedom for such a problem are $v = n - m$, where n is the sample size and m is the number of variables, both dependent and independent. Thus, for a sample value of $R = 0.42$ based on a sample of 50 items and measurements of 4 variables (one dependent plus three independent), one would conclude that it is significant at $P = 0.05$ but not at $P = 0.01$. The appropriate degrees of freedom are $50 - 4 = 46$, requiring interpolation, but since these conclusions are true for both $v = 45$ and $v = 50$, bracketing the correct value for the degrees of freedom, one need not interpolate. Further details are given in Section 16.4.

This table is reproduced by permission from George W. Snedecor, *Statistical Methods,* 5th ed., © 1956, by The Iowa State University Press.

TABLE **R** Critical values for correlation coefficients

ν	α	k 1	2	3	4	ν	α	k 1	2	3	4
		Number of independent variables						Number of independent variables			
1	.05	.997	.999	.999	.999	24	.05	.388	.470	.523	.562
	.01	1.000	1.000	1.000	1.000		.01	.496	.565	.609	.642
2	.05	.950	.975	.983	.987	25	.05	.381	.462	.514	.553
	.01	.990	.995	.997	.998		.01	.487	.555	.600	.633
3	.05	.878	.930	.950	.961	26	.05	.374	.454	.506	.545
	.01	.959	.976	.983	.987		.01	.478	.546	.590	.624
4	.05	.811	.881	.912	.930	27	.05	.367	.446	.498	.536
	.01	.917	.949	.962	.970		.01	.470	.538	.582	.615
5	.05	.754	.836	.874	.898	28	.05	.361	.439	.490	.529
	.01	.874	.917	.937	.949		.01	.463	.530	.573	.606
6	.05	.707	.795	.839	.867	29	.05	.355	.432	.482	.521
	.01	.834	.886	.911	.927		.01	.456	.522	.565	.598
7	.05	.666	.758	.807	.838	30	.05	.349	.426	.476	.514
	.01	.798	.855	.885	.904		.01	.449	.514	.558	.591
8	.05	.632	.726	.777	.811	35	.05	.325	.397	.445	.482
	.01	.765	.827	.860	.882		.01	.418	.481	.523	.556
9	.05	.602	.697	.750	.786	40	.05	.304	.373	.419	.455
	.01	.735	.800	.836	.861		.01	.393	.454	.494	.526
10	.05	.576	.671	.726	.763	45	.05	.288	.353	.397	.432
	.01	.708	.776	.814	.840		.01	.372	.430	.470	.501
11	.05	.553	.648	.703	.741	50	.05	.273	.336	.379	.412
	.01	.684	.753	.793	.821		.01	.354	.410	.449	.479
12	.05	.532	.627	.683	.722	60	.05	.250	.308	.348	.380
	.01	.661	.732	.773	.802		.01	.325	.377	.414	.442
13	.05	.514	.608	.664	.703	70	.05	.232	.286	.324	.354
	.01	.641	.712	.755	.785		.01	.302	.351	.386	.413
14	.05	.497	.590	.646	.686	80	.05	.217	.269	.304	.332
	.01	.623	.694	.737	.768		.01	.283	.330	.362	.389
15	.05	.482	.574	.630	.670	90	.05	.205	.254	.288	.315
	.01	.606	.677	.721	.752		.01	.267	.312	.343	.368
16	.05	.468	.559	.615	.655	100	.05	.195	.241	.274	.300
	.01	.590	.662	.706	.738		.01	.254	.297	.327	.351
17	.05	.456	.545	.601	.641	125	.05	.174	.216	.246	.269
	.01	.575	.647	.691	.724		.01	.228	.266	.294	.316
18	.05	.444	.532	.587	.628	150	.05	.159	.198	.225	.247
	.01	.561	.633	.678	.710		.01	.208	.244	.270	.290
19	.05	.433	.520	.575	.615	200	.05	.138	.172	.196	.215
	.01	.549	.620	.665	.698		.01	.181	.212	.234	.253
20	.05	.423	.509	.563	.604	300	.05	.113	.141	.160	.176
	.01	.537	.608	.652	.685		.01	.148	.174	.192	.208
21	.05	.413	.498	.522	.592	400	.05	.098	.122	.139	.153
	.01	.526	.596	.641	.674		.01	.128	.151	.167	.180
22	.05	.404	.488	.542	.582	500	.05	.088	.109	.124	.137
	.01	.515	.585	.630	.663		.01	.115	.135	.150	.162
23	.05	.396	.479	.532	.572	1,000	.05	.062	.077	.088	.097
	.01	.505	.574	.619	.652		.01	.081	.096	.106	.115

TABLE **S** Critical values for Kendall's rank correlation
coefficient τ

This table furnishes 0.10, 0.05, and 0.01 critical values for Kendall's rank correlation coefficient τ. These values are given for every sample size between $n = 4$ and $n = 40$. The probabilities are based on a two-tailed test. When a one-tailed test is desired, halve the probabilities at the head of the columns.

The table is used to test the null hypothesis that the variates in two samples are arrayed at random with respect to each other, that is, that the parametric rank correlation coefficient is zero.

To test the significance of a correlation coefficient, enter the table at the appropriate sample size and find the critical value for the chosen probability. For example, for a sample size of 15 the 5% critical value of $\tau = 0.390$. Thus, an observed value of 0.498 would be considered significant at the 5% but not at the 1% level. Negative correlations are considered as positive for purposes of this test. For sample sizes $n > 40$ use the asymptotic approximation given in Box 15.7.

This table is used to test the statistical significance of Kendall's rank correlation coefficients (Section 15.8).

The values in this table were derived from those in table XI of J. V. Bradley, *Distribution-Free Statistical Tests* (Prentice-Hall, New Jersey, 1968).

TABLE S Critical values for Kendall's rank correlation coefficient τ

$n \backslash \alpha$	0.10	0.05	0.01
4	1.000	-	-
5	0.800	1.000	-
6	0.733	0.867	1.000
7	0.619	0.714	0.905
8	0.571	0.643	0.786
9	0.500	0.556	0.722
10	0.467	0.511	0.644
11	0.418	0.491	0.600
12	0.394	0.455	0.576
13	0.359	0.436	0.564
14	0.363	0.407	0.516
15	0.333	0.390	0.505
16	0.317	0.383	0.483
17	0.309	0.368	0.471
18	0.294	0.346	0.451
19	0.287	0.333	0.439
20	0.274	0.326	0.421
21	0.267	0.314	0.410
22	0.264	0.307	0.394
23	0.257	0.296	0.391
24	0.246	0.290	0.377
25	0.240	0.287	0.367
26	0.237	0.280	0.360
27	0.231	0.271	0.356
28	0.228	0.265	0.344
29	0.222	0.261	0.340
30	0.218	0.255	0.333
31	0.213	0.252	0.325
32	0.210	0.246	0.323
33	0.205	0.242	0.314
34	0.201	0.237	0.312
35	0.197	0.234	0.304
36	0.194	0.232	0.302
37	0.192	0.228	0.297
38	0.189	0.223	0.292
39	0.188	0.220	0.287
40	0.185	0.218	0.285

TABLE T   Critical values of Olmstead and Tukey's test criterion

This table furnishes conservative critical values for the absolute value of the quadrant sum employed by Olmstead and Tukey as a test criterion for their corner test of association. The technique for obtaining a quadrant sum is explained in Section 15.8, and the significance level of a given sum is furnished in this table. Where two magnitudes of the quadrant sums are furnished for any significance level, the smaller magnitude applies to large samples, the larger magnitude to small samples. The probabilities given in this table are not accurate when the quadrant sum $\geq (2n - 6)$ and n is small. For sample sizes ≥ 14 the critical values up to $\alpha = 0.01$ are satisfactory.

This table is taken from P.S. Olmstead and J. W. Tukey (*Ann. Math. Stat.* **18:**496–513, 1947) with permission of the publisher.

α	Absolute value of quadrant sum
.10	9
.05	11
.02	13
.01	14, 15
.005	15, 17
.002	17, 19
.001	18, 21

TABLE U   Critical values of U, the Mann–Whitney statistic

The quantity U is known as the Wilcoxon two-sample statistic or as the Mann–Whitney statistic. Critical values are tabulated for two samples of sizes n_1 and n_2, where $n_1 \geq n_2$, up to $n_1 = n_2 = 20$. As presented here (and discussed in Section 13.11), the upper bounds of the critical values are such that the sample statistic U_s has to be greater than a given critical value. Some tables of these critical values give the lower bounds. The probabilities at the heads of the columns are based on a one-tailed test and represent the proportion of the area of the distribution of U in one tail beyond the critical value. The following one-tailed probabilities are furnished: 0.10, 0.05, 0.025, 0.01, 0.005, and 0.001. For a two-tailed test use the same critical values but double the probability at the heads of the columns.

We find the critical value of U ($\alpha = 0.025$, one-tailed) for two samples $n_1 = 14$, $n_2 = 12$ to be 123. Any value of $U_s > 123$ will be significant at $\alpha < 0.025$. When $n_2 > 20$, or when there are tied values across both samples, the significance of U_s can be tested by a formula given in Box 13.7.

This table is useful for significance testing in the Mann–Whitney U-test and the Wilcoxon two-sample test (Section 13.11 and Box 13.7), both of which are nonparametric tests of differences between two samples.

This table was extracted from a more extensive one (table 11.4) in D. B. Owen, *Handbook of Statistical Tables* (Addison-Wesley, Reading, Mass., 1962) with permission of the publishers.

TABLE U Critical values of U, the Mann–Whitney statistic

n_1	n_2	0.10	0.05	0.025	0.01	0.005	0.001
3	2	6					
	3	8	9				
4	2	8					
	3	11	12				
	4	13	15	16			
5	2	9	10				
	3	13	14	15			
	4	16	18	19	20		
	5	20	21	23	24	25	
6	2	11	12				
	3	15	16	17			
	4	19	21	22	23	24	
	5	23	25	27	28	29	
	6	27	29	31	33	34	
7	2	13	14				
	3	17	19	20	21		
	4	22	24	25	27	28	
	5	27	29	30	32	34	
	6	31	34	36	38	39	42
	7	36	38	41	43	45	48
8	2	14	15	16			
	3	19	21	22	24		
	4	25	27	28	30	31	
	5	30	32	34	36	38	40
	6	35	38	40	42	44	47
	7	40	43	46	49	50	54
	8	45	49	51	55	57	60
9	1	9					
	2	16	17	18			
	3	22	23	25	26	27	
	4	27	30	32	33	35	
	5	33	36	38	40	42	44
	6	39	42	44	47	49	52
	7	45	48	51	54	56	60
	8	50	54	57	61	63	67
	9	56	60	64	67	70	74
10	1	10					
	2	17	19	20			
	3	24	26	27	29	30	
	4	30	33	35	37	38	40
	5	37	39	42	44	46	49
	6	43	46	49	52	54	57
	7	49	53	56	59	61	65
	8	56	60	63	67	69	74
	9	62	66	70	74	77	82
	10	68	73	77	81	84	90

ABLE U   Critical values of U, the Mann–Whitney statistic (*continued*)

n_1	n_2	α 0.10	0.05	0.025	0.01	0.005	0.001
11	1	11					
	2	19	21	22			
	3	26	28	30	32	33	
	4	33	36	38	40	42	44
	5	40	43	46	48	50	53
	6	47	50	53	57	59	62
	7	54	58	61	65	67	71
	8	61	65	69	73	75	80
	9	68	72	76	81	83	89
	10	74	79	84	88	92	98
	11	81	87	91	96	100	106
12	1	12					
	2	20	22	23			
	3	28	31	32	34	35	
	4	36	39	41	42	45	48
	5	43	47	49	52	54	58
	6	51	55	58	61	63	68
	7	58	63	66	70	72	77
	8	66	70	74	79	81	87
	9	73	78	82	87	90	96
	10	81	86	91	96	99	106
	11	88	94	99	104	108	115
	12	95	102	107	113	117	124
13	1	13					
	2	22	24	25	26		
	3	30	33	35	37	38	
	4	39	42	44	47	49	51
	5	47	50	53	56	58	62
	6	55	59	62	66	68	73
	7	63	67	71	75	78	83
	8	71	76	80	84	87	93
	9	79	84	89	94	97	103
	10	87	93	97	103	106	113
	11	95	101	106	112	116	123
	12	103	109	115	121	125	133
	13	111	118	124	130	135	143
14	1	14					
	2	24	25	27	28		
	3	32	35	37	40	41	
	4	41	45	47	50	52	55
	5	50	54	57	60	63	67
	6	59	63	67	71	73	78
	7	67	72	76	81	83	89
	8	76	81	86	90	94	100
	9	85	90	95	100	104	111
	10	93	99	104	110	114	121
	11	102	108	114	120	124	132
	12	110	117	123	130	134	143
	13	119	126	132	139	144	153
	14	127	135	141	149	154	164

TABLE U Critical values of U, the Mann–Whitney statistic (*continued*

n_1	n_2	0.10	0.05	0.025	0.01	0.005	0.001
15	1	15					
	2	25	27	29	30		
	3	35	38	40	42	43	
	4	44	48	50	53	55	59
	5	53	57	61	64	67	71
	6	63	67	71	75	78	83
	7	72	77	81	86	89	95
	8	81	87	91	96	100	106
	9	90	96	101	107	111	118
	10	99	106	111	117	121	129
	11	108	115	121	128	132	141
	12	117	125	131	138	143	152
	13	127	134	141	148	153	163
	14	136	144	151	159	164	174
	15	145	153	161	169	174	185
16	1	16					
	2	27	29	31	32		
	3	37	40	42	45	46	
	4	47	50	53	57	59	62
	5	57	61	65	68	71	75
	6	67	71	75	80	83	88
	7	76	82	86	91	94	101
	8	86	92	97	102	106	113
	9	96	102	107	113	117	125
	10	106	112	118	124	129	137
	11	115	122	129	135	140	149
	12	125	132	139	146	151	161
	13	134	143	149	157	163	173
	14	144	153	160	168	174	185
	15	154	163	170	179	185	197
	16	163	173	181	190	196	208
17	1	17					
	2	28	31	32	34		
	3	39	42	45	47	49	51
	4	50	53	57	60	62	66
	5	60	65	68	72	75	80
	6	71	76	80	84	87	93
	7	81	86	91	96	100	106
	8	91	97	102	108	112	119
	9	101	108	114	120	124	132
	10	112	119	125	132	136	145
	11	122	130	136	143	148	158
	12	132	140	147	155	160	170
	13	142	151	158	166	172	183
	14	153	161	169	178	184	195
	15	163	172	180	189	195	208
	16	173	183	191	201	207	220
	17	183	193	202	212	219	232

TABLE U   Critical values of U, the Mann–Whitney statistic (*continued*)

n_1	n_2	0.10	0.05	0.025	0.01	0.005	0.001
18	1	18					
	2	30	32	34	36		
	3	41	45	47	50	52	54
	4	52	56	60	63	66	69
	5	63	68	72	76	79	84
	6	74	80	84	89	92	98
	7	85	91	96	102	105	112
	8	96	103	108	114	118	126
	9	107	114	120	126	131	139
	10	118	125	132	139	143	153
	11	129	137	143	151	156	166
	12	139	148	155	163	169	179
	13	150	159	167	175	181	192
	14	161	170	178	187	194	206
	15	172	182	190	200	206	219
	16	182	193	202	212	218	232
	17	193	204	213	224	231	245
	18	204	215	225	236	243	258
19	1	18	19				
	2	31	34	36	37	38	
	3	43	47	50	53	54	57
	4	55	59	63	67	69	73
	5	67	72	76	80	83	88
	6	78	84	89	94	97	103
	7	90	96	101	107	111	118
	8	101	108	114	120	124	132
	9	113	120	126	133	138	146
	10	124	132	138	146	151	161
	11	136	144	151	159	164	175
	12	147	156	163	172	177	188
	13	158	167	175	184	190	202
	14	169	179	188	197	203	216
	15	181	191	200	210	216	230
	16	192	203	212	222	230	244
	17	203	214	224	235	242	257
	18	214	226	236	248	255	271
	19	226	238	248	260	268	284

TABLE U Critical values of U, the Mann–Whitney statistic (*continued*

n_1	n_2	α					
		0.10	0.05	0.025	0.01	0.005	0.001
20	1	19	20				
	2	33	36	38	39	40	
	3	45	49	52	55	57	60
	4	58	62	66	70	72	77
	5	70	75	80	84	87	93
	6	82	88	93	98	102	108
	7	94	101	106	112	116	124
	8	106	113	119	126	130	139
	9	118	126	132	140	144	154
	10	130	138	145	153	158	168
	11	142	151	158	167	172	183
	12	154	163	171	180	186	198
	13	166	176	184	193	200	212
	14	178	188	197	207	213	226
	15	190	200	210	220	227	241
	16	201	213	222	233	241	255
	17	213	225	235	247	254	270
	18	225	237	248	260	268	284
	19	237	250	261	273	281	298
	20	249	262	273	286	295	312

TABLE V Critical values of the Wilcoxon rank sum

This table furnishes critical values for the one-tailed test of significance of the rank sum T_s obtained in Wilcoxon's matched-pairs signed-ranks test. Critical values are given for one-tailed probabilities 0.05, 0.025, 0.01, and 0.005 and for sample sizes from $n = 5$ to $n = 50$. Since the exact probability level desired cannot be obtained with integral critical values of T, two such values and their attendant probabilities bracketing the desired significance level are furnished. Thus, to find the significant 1% values for $n = 19$, we note the two critical values of T, 37 and 38, in the table. The probabilities corresponding to these two values of T are 0.0090 and 0.0102. Clearly a rank sum of $T_s = 37$ would have a probability of less than 0.01 and would be considered significant by the stated criterion. For two-tailed tests in which the alternative hypothesis is that the pairs could differ in either direction, double the probabilities stated at the head of the table. For two-tailed tests with sample sizes $n > 50$ compute

$$t_{\alpha[\infty]} = \left[T_s - \frac{n(n + 1)}{4} \right] \Bigg/ \sqrt{\frac{n(n + 1)(2n + 1)}{24}}$$

This table is used for Wilcoxon's matched-pairs signed-ranks test (Section 13.12).

The table was prepared on a computer using a recursion equation given in D. B. Owen, *Handbook of Statistical Tables* (Addison-Wesley, Reading, Mass., 1962, p. 325).

$$C = n_1 n_2 + \frac{n_2(n_2 + 1)}{2} - \sum R_i$$

C or $n_1 n_2 - C$ larger $= U_s$

TABLE V Critical values of the Wilcoxon rank sum

Nominal α

n	0.05 T	α	0.025 T	α	0.01 T	α	0.005 T	α
5	0	.0312						
	1	.0625						
6	2	.0469	0	.0156				
	3	.0781	1	.0312				
7	3	.0391	2	.0234	0	.0078		
	4	.0547	3	.0391	1	.0156		
8	5	.0391	3	.0195	1	.0078	0	.0039
	6	.0547	4	.0273	2	.0117	1	.0078
9	8	.0488	5	.0195	3	.0098	1	.0039
	9	.0645	6	.0273	4	.0137	2	.0059
10	10	.0420	8	.0244	5	.0098	3	.0049
	11	.0527	9	.0322	6	.0137	4	.0068
11	13	.0415	10	.0210	7	.0093	5	.0049
	14	.0508	11	.0269	8	.0122	6	.0068
12	17	.0461	13	.0212	9	.0081	7	.0046
	18	.0549	14	.0261	10	.0105	8	.0061
13	21	.0471	17	.0239	12	.0085	9	.0040
	22	.0549	18	.0287	13	.0107	10	.0052
14	25	.0453	21	.0247	15	.0083	12	.0043
	26	.0520	22	.0290	16	.0101	13	.0054
15	30	.0473	25	.0240	19	.0090	15	.0042
	31	.0535	26	.0277	20	.0108	16	.0051
16	35	.0467	29	.0222	23	.0091	19	.0046
	36	.0523	30	.0253	24	.0107	20	.0055
17	41	.0492	34	.0224	27	.0087	23	.0047
	42	.0544	35	.0253	28	.0101	24	.0055
18	47	.0494	40	.0241	32	.0091	27	.0045
	48	.0542	41	.0269	33	.0104	28	.0052
19	53	.0478	46	.0247	37	.0090	32	.0047
	54	.0521	47	.0273	38	.0102	33	.0054
20	60	.0487	52	.0242	43	.0096	37	.0047
	61	.0527	53	.0266	44	.0107	38	.0053

TABLE V Critical values of the Wilcoxon rank sum (*continued*)

	Nominal α							
	0.05		**0.025**		**0.01**		**0.005**	
n	T	α	T	α	T	α	T	α
21	67	.0479	58	.0230	49	.0097	42	.0045
	68	.0516	59	.0251	50	.0108	43	.0051
22	75	.0492	65	.0231	55	.0095	48	.0046
	76	.0527	66	.0250	56	.0104	49	.0052
23	83	.0490	73	.0242	62	.0098	54	.0046
	84	.0523	74	.0261	63	.0107	55	.0051
24	91	.0475	81	.0245	69	.0097	61	.0048
	92	.0505	82	.0263	70	.0106	62	.0053
25	100	.0479	89	.0241	76	.0094	68	.0048
	101	.0507	90	.0258	77	.0101	69	.0053
26	110	.0497	98	.0247	84	.0095	75	.0047
	111	.0524	99	.0263	85	.0102	76	.0051
27	119	.0477	107	.0246	92	.0093	83	.0048
	120	.0502	108	.0260	93	.0100	84	.0052
28	130	.0496	116	.0239	101	.0096	91	.0048
	131	.0521	117	.0252	102	.0102	92	.0051
29	140	.0482	126	.0240	110	.0095	100	.0049
	141	.0504	127	.0253	111	.0101	101	.0053
30	151	.0481	137	.0249	120	.0098	109	.0050
	152	.0502	138	.0261	121	.0104	110	.0053
31	163	.0491	147	.0239	130	.0099	118	.0049
	164	.0512	148	.0251	131	.0105	119	.0052
32	175	.0492	159	.0249	140	.0097	128	.0050
	176	.0512	160	.0260	141	.0103	129	.0053
33	187	.0485	170	.0242	151	.0099	138	.0049
	188	.0503	171	.0253	152	.0104	139	.0052
34	200	.0488	182	.0242	162	.0098	148	.0048
	201	.0506	183	.0252	163	.0103	149	.0051
35	213	.0484	195	.0247	173	.0096	159	.0048
	214	.0501	196	.0257	174	.0100	160	.0051

TABLE V Critical values of the Wilcoxon rank sum (*continued*)

Nominal α

n	0.05		0.025		0.01		0.005	
	T	α	*T*	α	*T*	α	*T*	α
36	227	.0489	208	.0248	185	.0096	171	.0050
	228	.0505	209	.0258	186	.0100	172	.0052
37	241	.0487	221	.0245	198	.0099	182	.0048
	242	.0503	222	.0254	199	.0103	183	.0050
38	256	.0493	235	.0247	211	.0099	194	.0048
	257	.0509	236	.0256	212	.0104	195	.0050
39	271	.0493	249	.0246	224	.0099	207	.0049
	272	.0507	250	.0254	225	.0103	208	.0051
40	286	.0486	264	.0249	238	.0100	220	.0049
	287	.0500	265	.0257	239	.0104	221	.0051
41	302	.0488	279	.0248	252	.0100	233	.0048
	303	.0501	280	.0256	253	.0103	234	.0050
42	319	.0496	294	.0245	266	.0098	247	.0049
	320	.0509	295	.0252	267	.0102	248	.0051
43	336	.0498	310	.0245	281	.0098	261	.0048
	337	.0511	311	.0252	282	.0102	262	.0050
44	353	.0495	327	.0250	296	.0097	276	.0049
	354	.0507	328	.0257	297	.0101	277	.0051
45	371	.0498	343	.0244	312	.0098	291	.0049
	372	.0510	344	.0251	313	.0101	292	.0051
46	389	.0497	361	.0249	328	.0098	307	.0050
	390	.0508	362	.0256	329	.0101	308	.0052
47	407	.0490	378	.0245	345	.0099	322	.0048
	408	.0501	379	.0251	346	.0102	323	.0050
48	426	.0490	396	.0244	362	.0099	339	.0050
	427	.0500	397	.0251	363	.0102	340	.0051
49	446	.0495	415	.0247	379	.0098	355	.0049
	447	.0505	416	.0253	380	.0100	356	.0050
50	466	.0495	434	.0247	397	.0098	373	.0050
	467	.0506	435	.0253	398	.0101	374	.0051

TABLE W   Critical values of the two-sample Kolmogorov–Smirnov statistic

This table furnishes *upper* critical values of $n_1 n_2 D$, the Kolmogorov–Smirnov test statistic D multiplied by the two sample sizes n_1 and n_2. In the two-sample case D is defined as the largest unsigned difference between the two relative cumulative frequency distributions representing the observed frequencies of the two samples. Critical values are tabulated for two samples of sizes n_1 and n_2 up to $n_1 = n_2 = 25$. Sample sizes n_1 are given at the left margin of the table; sample sizes n_2 are given across the top of the table at the heads of the columns. The six values furnished at the intersection of two sample sizes represent the following six two-tailed probabilities: 0.10, 0.05, 0.025, 0.01, 0.005, and 0.001.

For the sample statistic $n_1 n_2 D$ to be significant, it has to equal or exceed a tabled critical value.

For the samples $n_1 = 16$ and $n_2 = 10$, the 5% critical value of $n_1 n_2 D$ is 84. Any value of $n_1 n_2 D \geq 84$ will be significant at $P \leq 0.05$. When either n_1 or n_2 are > 25, approximate tests can be carried out as shown in Box 13.9.

This table is used for significance testing in the two-sample Kolmogorov–Smirnov test (Section 13.11 and Box 13.9) in nonparametric tests for the difference between two sample distributions.

When a one-sided test is desired, approximate probabilities can be obtained from this table by doubling the nominal α values. However, these are not exact because the distribution of cumulative frequencies is discrete. A one-sided table is furnished by M. H. Gail and S. B. Green (*J. Am. Stat. Assn.* **71**:757–760, 1976).

This table was copied from table 55 in E. S. Pearson and H. O. Hartley, *Biometrika Tables for Statisticians*, Vol. 2 (Cambridge University Press, 1972) with permission of the publisher.

TABLE W Critical values of the two-sample Kolmogorov–Smirnov statistic

n_1	α	1	2	3	4	5	6	7	8	9	10	11	12	13	14	15	16	17	18	19	20	21	22	23	24	25
1	.1	-	-	-	-	-	-	-	-	-	-	-	-	-	-	-	-	-	-	19	20	21	22	23	24	25
	.05	-	-	-	-	-	-	-	-	-	-	-	-	-	-	-	-	-	-	-	-	-	-	-	-	-
	.025	-	-	-	-	-	-	-	-	-	-	-	-	-	-	-	-	-	-	-	-	-	-	-	-	-
	.01	-	-	-	-	-	-	-	-	-	-	-	-	-	-	-	-	-	-	-	-	-	-	-	-	-
	.005	-	-	-	-	-	-	-	-	-	-	-	-	-	-	-	-	-	-	-	-	-	-	-	-	-
	.001	-	-	-	-	-	-	-	-	-	-	-	-	-	-	-	-	-	-	-	-	-	-	-	-	-
2	.1	-	-	-	-	10	12	14	16	18	18	20	22	24	24	26	28	30	32	32	34	36	38	38	40	42
	.05	-	-	-	-	-	12	-	16	18	20	22	24	26	26	28	30	32	34	36	38	38	40	42	44	46
	.025	-	-	-	-	-	-	-	-	-	-	-	24	26	28	30	32	34	36	38	40	40	42	44	46	48
	.01	-	-	-	-	-	-	-	-	-	-	-	-	-	-	-	-	-	-	38	40	42	44	46	48	-
	.005	-	-	-	-	-	-	-	-	-	-	-	-	-	-	-	-	-	-	-	-	-	-	-	-	50
	.001	-	-	-	-	-	-	-	-	-	-	-	-	-	-	-	-	-	-	-	-	-	-	-	-	-
3	.1	-	-	9	-	15	15	18	21	21	24	27	27	30	33	33	36	36	39	42	42	45	48	48	51	54
	.05	-	-	-	-	15	18	21	21	24	27	30	30	33	36	36	39	42	45	45	48	51	51	54	57	60
	.025	-	-	-	-	-	18	21	24	27	30	30	33	36	39	39	42	45	48	51	51	54	57	60	60	63
	.01	-	-	-	-	-	-	-	-	27	30	33	36	39	42	42	45	48	51	54	57	57	60	63	66	69
	.005	-	-	-	-	-	-	-	-	-	-	-	36	39	42	45	48	51	54	57	57	60	63	66	69	72
	.001	-	-	-	-	-	-	-	-	-	-	-	-	-	-	-	-	-	-	-	-	63	66	69	72	75
4	.1	-	-	12	16	16	18	21	24	27	28	29	36	35	38	40	44	44	46	49	52	52	56	57	60	63
	.05	-	-	-	16	20	20	24	28	28	30	33	36	39	42	44	48	48	50	53	60	59	62	64	68	68
	.025	-	-	-	-	20	24	28	28	32	36	36	40	44	44	45	52	52	54	57	64	63	66	69	72	75
	.01	-	-	-	-	24	24	28	32	36	36	40	44	48	48	52	56	60	60	64	68	72	72	76	80	84
	.005	-	-	-	-	-	-	-	32	36	40	44	48	48	52	56	60	64	64	68	72	76	76	80	84	88
	.001	-	-	-	-	-	-	-	-	-	-	-	-	52	56	60	64	68	72	76	76	80	84	88	92	96

n_2

n_2

n_1	α	1	2	3	4	5	6	7	8	9	10	11	12	13	14	15	16	17	18	19	20	21	22	23	24	25
5	.1	–	10	15	16	20	24	25	27	30	35	35	36	40	42	50	48	50	52	56	60	60	63	65	67	75
	.05	–	–	15	20	25	24	28	30	35	40	39	43	45	46	55	54	55	60	61	65	69	70	72	76	80
	.025	–	–	–	20	25	30	30	32	36	40	44	45	47	51	55	59	60	65	66	75	74	78	80	81	90
	.01	–	–	–	–	25	30	35	35	40	45	45	50	52	56	60	64	68	70	71	80	80	83	87	90	95
	.005	–	–	–	–	–	–	35	40	45	45	50	55	55	60	65	70	70	72	76	85	84	88	92	95	100
	.001	–	–	–	–	–	–	–	–	45	50	55	60	65	70	70	75	80	85	85	90	95	100	105	105	110
6	.1	–	12	15	18	24	30	28	30	33	36	38	48	46	48	51	54	56	66	64	66	69	70	73	78	78
	.05	–	–	18	20	24	30	30	34	39	40	43	48	52	54	57	60	62	72	70	72	75	78	80	90	88
	.025	–	–	18	24	30	36	35	36	42	44	48	54	54	58	63	64	67	78	76	78	81	86	86	96	96
	.01	–	–	–	24	30	36	42	40	48	48	54	60	60	64	69	72	73	84	83	88	90	92	97	102	107
	.005	–	–	–	–	30	36	42	42	48	50	54	60	65	66	72	74	79	84	89	90	96	98	103	108	113
	.001	–	–	–	–	–	–	–	48	54	60	66	66	72	78	84	84	85	96	96	100	105	110	114	120	125
7	.1	–	14	18	21	25	28	35	34	36	40	44	46	50	56	56	59	61	65	69	72	77	77	80	84	86
	.05	–	–	21	24	28	30	42	40	42	46	48	53	56	63	62	64	68	72	76	79	91	84	89	92	97
	.025	–	–	21	28	30	35	42	41	45	49	52	56	58	70	68	73	77	80	84	86	98	96	98	102	105
	.01	–	–	–	28	35	36	49	48	49	53	59	60	65	77	75	77	84	87	91	93	105	103	108	112	115
	.005	–	–	–	–	35	42	49	48	54	56	63	65	70	77	77	84	85	91	95	99	112	110	112	119	122
	.001	–	–	–	–	–	–	–	56	63	63	70	72	78	84	90	96	98	101	107	112	119	125	126	133	136
8	.1	–	16	21	24	27	30	34	40	40	44	48	52	54	58	60	72	68	72	74	80	81	84	89	96	95
	.05	–	16	21	28	30	34	40	48	46	48	53	60	62	64	67	80	77	80	82	88	89	94	98	104	104
	.025	–	24	24	28	32	36	41	48	48	54	58	64	65	70	74	80	80	86	90	96	97	102	106	112	112
	.01	–	–	–	32	35	40	48	56	55	60	64	68	72	76	81	88	88	94	98	104	107	112	115	128	125
	.005	–	–	–	32	40	42	48	56	56	62	66	72	78	82	88	96	96	100	104	112	115	120	122	136	134
	.001	–	–	–	–	–	48	56	64	64	70	77	80	88	90	97	104	111	112	117	124	126	132	137	152	150

TABLE W Critical values of the two-sample Kolmogorov–Smirnov statistic (*continued*)

n_2

n_1	α	1	2	3	4	5	6	7	8	9	10	11	12	13	14	15	16	17	18	19	20	21	22	23	24	25
9	.1	-	18	21	27	30	33	36	40	54	50	52	57	59	63	69	69	74	81	80	84	90	91	94	99	101
	.05	-	18	24	28	35	39	42	46	54	53	59	63	65	70	75	78	82	90	89	93	99	101	106	111	114
	.025	-	-	27	32	36	42	45	48	63	60	63	69	72	76	81	85	90	99	98	100	108	110	115	120	123
	.01	-	-	27	36	40	45	49	55	63	63	70	75	78	84	90	94	99	108	107	111	117	122	126	132	135
	.005	-	-	-	36	45	48	54	56	72	70	72	78	82	89	93	99	102	117	114	117	123	127	134	138	144
	.001	-	-	-	-	45	54	63	64	72	80	81	87	91	98	105	110	117	126	126	133	138	144	152	156	162
10	.1	-	18	24	28	35	36	40	44	50	60	57	60	64	68	75	76	79	82	85	100	95	98	101	106	110
	.05	-	20	27	30	40	40	46	48	53	70	60	66	70	74	80	84	89	92	94	110	105	108	114	118	125
	.025	-	-	30	36	40	44	49	54	60	70	68	72	77	82	90	90	96	100	103	120	116	113	124	128	135
	.01	-	-	30	36	44	48	53	60	70	70	77	80	84	90	100	100	106	108	113	130	126	130	137	140	150
	.005	-	-	-	40	45	50	56	62	70	80	79	84	90	96	105	108	110	116	122	140	130	138	144	148	155
	.001	-	-	-	-	50	60	63	70	80	90	89	96	100	106	115	118	126	132	133	150	149	154	160	166	175
11	.1	-	20	27	29	35	38	44	48	52	57	66	64	67	73	76	80	85	88	92	96	101	110	108	111	117
	.05	-	22	30	33	39	43	48	53	59	60	77	72	75	82	84	89	93	97	102	107	112	121	119	124	129
	.025	-	-	30	36	44	48	52	58	63	68	77	76	84	87	94	96	102	107	111	116	123	132	131	137	140
	.01	-	-	33	40	45	54	59	64	70	77	88	86	91	96	102	106	110	118	122	127	134	143	142	150	154
	.005	-	-	-	44	50	55	63	66	72	79	88	88	97	101	109	112	119	125	130	136	143	154	153	159	164
	.001	-	-	-	-	55	66	70	77	81	89	99	99	108	115	120	127	132	140	146	154	157	176	173	184	184
12	.1	-	22	27	36	36	48	46	52	57	60	64	72	71	78	84	88	90	96	99	104	108	110	113	132	120
	.05	-	24	30	36	43	48	53	60	63	66	72	84	81	86	93	96	100	108	108	116	120	124	125	144	138
	.025	-	24	33	40	44	54	60	64	69	72	76	96	84	93	99	104	108	120	120	124	129	134	137	156	150
	.01	-	-	36	44	50	60	60	68	75	80	86	96	95	104	108	116	119	126	130	140	141	148	149	168	165
	.005	-	-	36	48	55	60	65	72	78	84	88	108	104	108	117	124	127	138	140	148	150	154	160	180	175
	.001	-	-	-	-	60	66	72	80	87	96	99	120	117	120	129	136	141	150	156	164	168	174	182	192	192

n_2

n_1	α	1	2	3	4	5	6	7	8	9	10	11	12	13	14	15	16	17	18	19	20	21	22	23	24	25
13	.1	-	24	30	35	40	46	50	54	59	64	67	71	91	78	87	91	96	99	104	108	113	117	120	125	131
	.05	-	26	33	39	45	52	56	62	65	70	75	81	91	89	96	101	105	110	114	120	126	130	135	140	145
	.025	-	26	36	44	47	54	58	65	72	77	84	84	104	100	104	111	114	120	126	130	137	141	146	151	158
	.01	-	-	39	48	52	60	65	72	78	84	91	95	117	104	115	121	127	131	138	143	150	156	161	166	172
	.005	-	-	39	48	55	65	70	78	82	90	97	104	117	115	122	128	135	141	145	154	161	168	171	177	184
	.001	-	-	-	52	65	72	78	88	91	100	108	117	130	129	137	143	152	156	164	169	179	185	191	199	200
14	.1	-	24	33	38	42	48	56	58	63	68	73	78	78	98	92	96	100	104	110	114	126	124	127	132	136
	.05	-	26	36	42	46	54	63	64	70	74	82	86	89	112	98	106	111	116	121	126	140	138	142	146	150
	.025	-	28	39	44	51	58	70	70	76	82	87	94	100	112	110	116	122	126	133	138	147	148	154	160	166
	.01	-	-	42	48	56	64	77	76	84	90	96	104	104	126	123	126	134	140	148	152	161	164	170	176	182
	.005	-	-	42	52	60	66	77	82	89	96	101	108	115	126	125	136	140	148	154	160	175	174	179	186	194
	.001	-	-	-	56	70	78	84	90	98	106	115	120	129	154	140	152	159	166	176	180	189	196	202	210	219
15	.1	-	26	33	40	50	51	56	60	69	75	76	84	87	92	105	101	105	111	114	125	126	130	134	141	145
	.05	-	28	36	44	55	57	62	67	75	80	84	93	96	98	120	114	116	123	127	135	138	144	149	156	160
	.025	-	30	39	45	55	63	68	74	81	90	94	99	104	110	135	119	129	135	141	150	153	154	163	168	175
	.01	-	-	42	52	60	69	75	81	90	100	102	108	115	123	135	133	142	147	152	160	168	173	179	186	195
	.005	-	-	45	56	65	72	77	88	93	105	109	117	122	125	150	144	148	156	161	170	177	182	187	198	205
	.001	-	-	-	60	70	84	90	97	105	115	120	129	137	140	165	162	165	174	180	195	198	205	210	222	230
16	.1	-	28	36	44	48	54	59	72	69	76	80	88	91	96	101	112	109	116	120	128	130	136	141	152	149
	.05	-	30	39	48	54	60	64	80	78	84	89	96	101	106	114	128	124	128	133	140	145	150	157	168	167
	.025	-	32	42	52	59	64	73	80	85	90	96	104	111	116	119	144	136	140	145	156	157	164	169	184	181
	.01	-	-	45	56	64	72	77	88	94	106	106	116	121	126	133	144	143	154	160	168	173	180	187	200	199
	.005	-	-	48	60	70	74	84	96	99	108	112	124	128	136	144	160	157	162	170	180	183	192	198	208	213
	.001	-	-	64	75	84	96	104	110	118	127	136	143	152	162	176	174	186	192	200	208	216	221	232	238	

TABLE W Critical values of the two-sample Kolmogorov–Smirnov statistic (*continued*)

n_1	α	1	2	3	4	5	6	7	8	9	10	11	12	13	14	15	16	17	18	19	20	21	22	23	24	25
17	.1	–	30	36	44	50	56	61	68	74	79	85	90	96	100	105	109	136	118	126	132	136	142	146	151	156
	.05	–	32	42	48	55	62	68	77	82	89	93	100	105	111	116	124	136	133	141	146	151	157	163	168	173
	.025	–	34	45	52	60	67	77	80	90	96	102	108	114	122	129	136	153	148	151	160	166	170	179	183	190
	.01	–	–	48	60	68	73	84	88	99	106	110	119	127	134	142	143	170	164	166	175	180	187	196	203	207
	.005	–	–	51	64	70	79	85	96	102	110	119	127	135	140	148	157	170	168	179	186	193	199	207	214	222
	.001	–	–	–	68	80	85	98	111	117	126	132	141	152	159	165	174	204	187	200	209	217	225	232	240	249
18	.1	–	32	39	46	52	66	65	72	81	82	88	96	99	104	111	116	118	144	133	136	144	148	152	162	162
	.05	–	34	45	50	60	72	72	80	90	92	97	108	110	116	123	128	133	162	142	152	159	164	170	180	180
	.025	–	36	48	54	65	78	80	86	99	100	107	120	120	126	135	140	148	162	159	166	174	178	184	198	196
	.01	–	–	51	60	70	84	87	94	108	108	118	126	131	140	147	154	164	180	176	182	189	196	204	216	216
	.005	–	–	54	64	72	84	91	100	117	116	125	138	141	148	156	162	168	198	180	194	201	208	216	228	231
	.001	–	–	–	72	85	96	101	112	126	132	140	150	156	166	174	186	187	216	212	214	225	234	242	252	257
19	.1	–	32	42	49	56	64	69	74	80	85	92	99	104	110	114	120	126	133	152	144	147	152	159	164	168
	.05	–	36	45	53	61	70	76	82	89	94	102	108	114	121	127	133	141	142	171	160	163	169	177	183	187
	.025	–	38	51	57	66	76	84	90	98	103	111	120	126	133	141	145	151	159	190	169	180	185	190	199	205
	.01	–	38	54	64	71	83	91	98	107	113	122	130	138	148	152	160	166	176	190	187	199	204	209	218	224
	.005	–	–	57	68	76	89	95	104	114	122	130	140	145	154	161	170	179	180	209	204	207	219	224	232	241
	.001	–	–	–	76	85	96	107	114	126	133	146	156	164	176	180	192	200	212	228	225	237	242	253	261	268
20	.1	–	34	42	52	60	66	72	80	84	100	96	104	108	114	125	128	132	136	144	160	154	160	164	172	180
	.05	–	38	48	60	65	72	79	88	93	110	107	116	120	126	135	140	146	152	160	180	173	176	184	192	200
	.025	–	40	51	64	75	78	86	96	100	116	116	124	130	138	150	156	160	166	169	200	180	192	199	208	215
	.01	–	40	57	68	80	88	93	104	111	130	127	140	143	152	160	168	175	182	187	220	199	212	219	228	235
	.005	–	–	57	72	85	90	99	112	117	130	136	148	154	160	170	180	186	194	204	220	217	226	233	244	250
	.001	–	–	–	76	90	100	112	124	133	150	154	164	169	180	195	200	209	214	225	260	239	254	262	272	280

n_2

n_1	α	1	2	3	4	5	6	7	8	9	10	11	12	13	14	15	16	17	18	19	20	21	22	23	24	25
21	.1	21	36	45	52	60	69	77	81	90	95	101	108	113	126	126	130	136	144	147	154	168	163	171	177	182
	.05	-	38	51	59	69	75	91	89	99	105	112	120	126	140	138	145	151	159	163	173	189	183	189	198	202
	.025	-	40	54	63	74	81	98	97	108	116	123	129	137	147	153	157	166	174	180	180	210	203	206	213	220
	.01	-	42	57	72	80	90	105	107	117	126	134	141	150	161	168	173	180	189	199	199	231	223	227	237	244
	.005	-	-	60	76	84	96	112	115	123	130	143	150	161	175	177	183	193	201	207	217	252	229	242	252	258
	.001	-	-	63	80	95	105	119	126	138	149	157	168	179	189	198	208	217	225	237	239	273	267	269	282	290
22	.1	22	38	48	56	63	70	77	84	91	98	110	110	117	124	130	136	142	148	152	160	163	198	173	182	189
	.05	-	40	51	62	70	78	84	94	101	108	121	124	130	138	144	150	157	164	169	176	183	198	194	204	209
	.025	-	42	57	66	78	86	96	102	110	118	132	134	141	148	154	164	170	178	185	192	203	220	214	222	228
	.01	-	44	60	72	83	92	103	112	122	130	143	148	156	164	173	180	187	196	204	212	223	242	237	242	250
	.005	-	-	63	76	88	98	110	120	127	138	154	154	168	174	182	192	199	208	219	226	229	264	253	258	268
	.001	-	-	66	84	100	110	125	132	144	154	176	174	185	196	205	216	225	234	242	254	267	286	282	292	299
23	.1	23	38	48	57	65	73	80	89	94	101	108	113	120	127	134	141	146	152	159	164	171	173	207	183	195
	.05	-	42	54	64	72	80	89	98	106	114	119	125	135	142	149	157	163	170	177	184	189	194	230	205	216
	.025	-	44	60	69	80	86	98	106	115	124	131	137	146	154	163	169	179	184	190	199	206	214	230	226	237
	.01	-	46	63	76	87	97	108	115	126	137	142	149	161	170	179	187	196	204	209	219	227	237	253	249	262
	.005	-	-	66	80	92	103	112	122	134	144	153	160	171	179	187	198	207	216	224	233	242	253	276	270	274
	.001	-	-	69	88	105	114	126	137	152	160	173	182	191	202	210	221	232	242	253	262	269	282	299	296	312
24	.1	24	40	51	60	67	78	84	96	99	106	111	132	125	132	141	152	151	162	164	172	177	182	183	216	204
	.05	-	44	57	68	76	90	92	104	111	118	124	144	140	146	156	168	168	180	183	192	198	204	205	240	225
	.025	-	46	60	72	81	96	102	112	120	128	137	156	151	160	168	184	183	198	199	208	213	222	226	264	238
	.01	-	48	66	80	90	102	112	128	132	140	150	168	166	176	186	200	203	216	218	228	237	242	249	288	262
	.005	-	-	69	84	95	108	119	136	138	148	159	180	177	186	198	208	214	228	232	244	252	258	270	288	283
	.001	-	-	72	92	105	120	133	152	156	166	176	192	199	210	222	232	240	252	261	272	282	292	296	336	312

TABLE W   Critical values of the two-sample Kolmogorov–Smirnov statistic (*continued*)

n_1	α	1	2	3	4	5	6	7	8	9	10	11	12	13	14	15	16	17	18	19	20	21	22	23	24	25
25	.1	25	42	54	63	75	78	86	95	101	110	117	120	131	136	145	149	156	162	168	180	182	189	195	204	225
	.05	-	46	60	68	80	88	97	104	114	125	129	138	145	150	160	167	173	180	187	200	202	209	216	225	250
	.025	-	48	63	75	90	96	105	112	123	135	140	150	158	166	175	181	190	196	205	215	220	228	237	238	275
	.01	-	50	69	84	95	107	115	125	135	150	154	165	172	182	195	199	207	216	224	235	244	250	262	262	300
	.005	-	-	72	88	100	113	122	134	144	155	164	175	184	194	205	213	222	231	241	250	258	268	274	283	325
	.001	-	-	75	96	110	125	136	150	162	175	184	192	200	219	230	238	249	257	268	280	290	299	312	312	350

TABLE X   Critical values of the δ-corrected one-sample KolmogorovSmirnov statistic

This table furnished critical values of the δ-corrected Kolmogorov–Smirnov test statistic, $g_{\delta,n}$, for testing goodness of fit of an observed cumulative frequency distribution to an expected distribution with specified parameters (that is, parameters not estimated from the data). Critical values of $g_{\delta,n}$ are given for three values of δ (0, 0.5, and 1) for every sample size from 3 to 100. The probabilities 0.2, 0.1, 0.05, 0.02, and 0.01 given at the heads of the columns of the table are the probabilities, α, that a sample value of g_δ equals or exceeds the tabulated value when the null hypothesis is true (sample drawn from a population with the expected distribution). The statistic g_δ is the maximum difference between an observed and an expected cumulative frequency distribution plotted at positions $F_\delta = (i - \delta)/(n - 2\delta + 1)$ for the ith observation.

Note that these critical values are not the same as those for tabulations of the traditional D_{max} Kolmogorov–Smirnov statistic. Critical values of D_{max} can be computed, however, as $g_{0.5,n} + 1/2n$. The new approach using δ = 0 and 1 is from H. J. Khamis (*J. Stat. Planning Inference* **24**:317–335, 1990). It provides a more powerful test.

For a sample of $n = 42$ items, the 5% critical values of g_0 and g_1 are 0.19050 and 0.19684, respectively. If sample values of either g_0 and g_1 are equal to or exceed their critical values, then the null hypothesis is rejected at the 5% level. For sample sizes $n > 100$, where the effect of δ is relatively unimportant, the critical value of $g_{0.5}$ can be approximated by

$$g_{0.5,n} = \sqrt{\frac{-\ln(\alpha/2)}{2n}} + \frac{1}{2n}$$

For $\alpha = 0.05$ this value is $1.35810/\sqrt{n} - 1/2n$, as shown in the last line of the table in the column for $\alpha = 0.05$.

The critical values of $g_{\delta,n}$ are used to test the deviations of an observed continuous frequency distribution from an expected one (Box 17.3, Section 17.2) and for setting confidence limits to a cumulative frequency distribution. The parameters of the expected distribution are *not* estimated from the observed distribution; that is, the observations are fitted to an extrinsic hypothesis. Appropriate critical values are provided in Table **Y** for the more common case of fitting a normal distribution to a sample using sample estimates of the mean and standard deviation.

This table is based on values provided by H. J. Khamis. A description of the table and of how it was computed was given in Khamis (*J. Stat. Planning Inference* **24**:317–335, 1990).

TABLE X Critical values of the δ-corrected one-sample Kolmogorov–Smirnov statistic

n	δ	α 0.2	0.1	0.05	0.02	0.01
3	0.0	.35477	.41811	.46702	.53456	.57900
	0.5	.39814	.46938	.54093	.61789	.66234
	1.0	.53584	.63160	.70760	.78456	.82900
4	0.0	.33435	.39075	.44641	.50495	.54210
	0.5	.36765	.44022	.49894	.56387	.60924
	1.0	.46154	.53829	.60468	.68377	.73409
5	0.0	.31556	.37359	.42174	.47692	.51576
	0.5	.34698	.40945	.46328	.52718	.56853
	1.0	.41172	.48153	.54273	.61133	.65692
6	0.0	.30244	.35522	.40045	.45440	.48988
	0.5	.32704	.38466	.43593	.49407	.53327
	1.0	.37706	.44074	.49569	.55969	.60287
7	0.0	.28991	.33905	.38294	.43337	.46761
	0.5	.31005	.36464	.41200	.46701	.50438
	1.0	.35066	.40892	.46010	.51968	.55970
8	0.0	.27828	.32538	.36697	.41522	.44819
	0.5	.29581	.34712	.39177	.44404	.47929
	1.0	.32925	.38365	.43160	.48732	.52519
9	0.0	.26794	.31325	.35277	.39922	.43071
	0.5	.28355	.33191	.37446	.42404	.45776
	1.0	.31157	.36287	.40794	.46067	.49652
10	0.0	.25884	.30221	.34022	.38481	.41517
	0.5	.27260	.31866	.35925	.40662	.43893
	1.0	.29668	.34525	.38798	.43809	.47220
11	0.0	.25071	.29227	.32894	.37187	.40122
	0.5	.26284	.30697	.34577	.39125	.42225
	1.0	.28388	.33008	.37084	.41864	.45127
12	0.0	.24325	.28330	.31869	.36019	.38856
	0.5	.25410	.29648	.33376	.37751	.40738
	1.0	.27269	.31686	.35588	.40167	.43298
13	0.0	.23639	.27515	.30935	.34954	.37703
	0.5	.24624	.28703	.32297	.36516	.39401
	1.0	.26279	.30520	.34265	.38668	.41680
14	0.0	.23010	.26767	.30081	.33980	.36649
	0.5	.23909	.27846	.31319	.35398	.38190
	1.0	.25395	.29478	.33086	.37331	.40238
15	0.0	.22430	.26077	.29296	.33083	.35679
	0.5	.23255	.27064	.30426	.34379	.37087
	1.0	.24600	.28541	.32026	.36128	.38940
16	0.0	.21895	.25439	.28570	.32256	.34784
	0.5	.22653	.26347	.29608	.33446	.36076
	1.0	.23879	.27692	.31065	.35039	.37764
17	0.0	.21397	.24847	.27897	.31489	.33953
	0.5	.22098	.25686	.28855	.32586	.35145
	1.0	.23221	.26918	.30189	.34045	.36691
18	0.0	.20933	.24296	.27270	.30775	.33181
	0.5	.21582	.25073	.28158	.31792	.34284
	1.0	.22617	.26208	.29386	.33134	.35707
19	0.0	.20498	.23781	.26685	.30108	.32459
	0.5	.21103	.24504	.27511	.31054	.33485
	1.0	.22060	.25553	.28646	.32295	.34801
20	0.0	.20089	.23298	.26137	.29484	.31784
	0.5	.20656	.23973	.26908	.30366	.32741
	1.0	.21544	.24947	.27961	.31518	.33962
21	0.0	.19705	.22844	.25622	.28898	.31149
	0.5	.20236	.23477	.26343	.29723	.32045
	1.0	.21064	.24384	.27325	.30796	.33182

TABLE **X** Critical values of the δ-corrected one-sample Kolmogorov–Smirnov statistic (*continued*)

				α		
n	δ	0.2	0.1	0.05	0.02	0.01
22	0.0	.19343	.22416	.25136	.28346	.30552
	0.5	.19843	.23011	.25814	.29121	.31393
	1.0	.20616	.23859	.26732	.30123	.32456
23	0.0	.19001	.22012	.24679	.27825	.29989
	0.5	.19472	.22572	.25317	.28554	.30780
	1.0	.20197	.23367	.26176	.29494	.31776
24	0.0	.18677	.21630	.24245	.27333	.29456
	0.5	.19121	.22159	.24847	.28021	.30202
	1.0	.19804	.22906	.25656	.28904	.31138
25	0.0	.18370	.21268	.23835	.26866	.28951
	0.5	.18790	.21768	.24404	.27516	.29657
	1.0	.19433	.22472	.25166	.28349	.30539
26	0.0	.18077	.20924	.23445	.26423	.28472
	0.5	.18476	.21397	.23984	.27039	.29140
	1.0	.19084	.22063	.24704	.27825	.29973
27	0.0	.17799	.20596	.23074	.26001	.28016
	0.5	.18178	.21046	.23586	.26586	.28650
	1.0	.18753	.21676	.24267	.27330	.29439
28	0.0	.17533	.20283	.22721	.25600	.27582
	0.5	.17894	.20712	.23208	.26156	.28185
	1.0	.18440	.21309	.23853	.26861	.28933
29	0.0	.17280	.19985	.22383	.25217	.27168
	0.5	.17624	.20393	.22847	.25747	.27742
	1.0	.18142	.20961	.23461	.26417	.28452
30	0.0	.17037	.19700	.22061	.24851	.26772
	0.5	.17365	.20090	.22504	.25356	.27320
	1.0	.17859	.20630	.23088	.25994	.27996
31	0.0	.16805	.19427	.21752	.24501	.26393
	0.5	.17119	.19800	.22176	.24983	.26917
	1.0	.17589	.20314	.22732	.25591	.27561
32	0.0	.16582	.19166	.21457	.24165	.26030
	0.5	.16882	.19522	.21862	.24627	.26531
	1.0	.17332	.20014	.22393	.25207	.27146
33	0.0	.16368	.18915	.21173	.23843	.25683
	0.5	.16656	.19256	.21561	.24286	.26162
	1.0	.17086	.19726	.22069	.24840	.26750
34	0.0	.16162	.18674	.20901	.23534	.25348
	0.5	.16439	.19001	.21273	.23958	.25808
	1.0	.16850	.19451	.21759	.24490	.26371
35	0.0	.15964	.18442	.20639	.23237	.25027
	0.5	.16230	.18756	.20996	.23644	.25469
	1.0	.16625	.19188	.21462	.24154	.26008
36	0.0	.15774	.18218	.20387	.22951	.24718
	0.5	.16029	.18521	.20730	.23343	.25143
	1.0	.16408	.18935	.21178	.23831	.25660
37	0.0	.15590	.18003	.20144	.22676	.24421
	0.5	.15836	.18294	.20474	.23052	.24829
	1.0	.16200	.18692	.20904	.23522	.25326
38	0.0	.15413	.17796	.19910	.22410	.24134
	0.5	.15650	.18076	.20228	.22773	.24527
	1.0	.16000	.18459	.20642	.23225	.25005
39	0.0	.15242	.17595	.19684	.22154	.23857
	0.5	.15471	.17866	.19991	.22504	.24236
	1.0	.15808	.18234	.20389	.22938	.24696
40	0.0	.15076	.17402	.19465	.21907	.23589
	0.5	.15297	.17663	.19762	.22244	.23955
	1.0	.15622	.18018	.20145	.22663	.24399

TABLE **X**   Critical values of the δ-corrected one-
sample Kolmogorov–Smirnov statistic
(*continued*)

n	δ	0.2	0.1	0.05	0.02	0.01
41	0.0	.14916	.17215	.19254	.21667	.23331
	0.5	.15130	.17467	.19540	.21993	.23684
	1.0	.15443	.17810	.19910	.22397	.24112
42	0.0	.14761	.17034	.19050	.21436	.23081
	0.5	.14968	.17278	.19327	.21751	.23422
	1.0	.15270	.17608	.19684	.22141	.23835
43	0.0	.14611	.16858	.18852	.21212	.22839
	0.5	.14811	.17094	.19120	.21517	.23169
	1.0	.15103	.17414	.19465	.21893	.23568
44	0.0	.14466	.16688	.18661	.20995	.22604
	0.5	.14659	.16917	.18920	.21290	.22924
	1.0	.14942	.17226	.19253	.21654	.23310
45	0.0	.14325	16524	.18475	.20785	.22377
	0.5	.14512	.16745	.18726	.21070	.22687
	1.0	.14786	.17044	.19049	.21423	.23060
46	0.0	.14188	.16364	.18295	.20581	.22157
	0.5	.14370	.16578	.18538	.20858	.22457
	1.0	.14635	.16868	.18851	.21199	.22818
47	0.0	.14055	.16208	.18120	.20383	.21943
	0.5	.14231	.16417	.18356	.20651	.22234
	1.0	.14488	.16697	.18659	.20982	.22584
48	0.0	.13926	.16058	.17950	.20190	.21735
	0.5	.14097	.16260	.18179	.20451	.22018
	1.0	.14346	.16532	.18473	.20772	.22357
49	0.0	.13800	.15911	.17785	.20003	.21534
	0.5	.13967	.16107	.18007	.20257	.21808
	1.0	.14208	.16371	.18293	.20568	.22137
50	0.0	.13678	.15769	.17624	.19822	.21337
	0.5	.13840	.15959	.17841	.20068	.21604
	1.0	.14074	.16216	.18117	.20370	.21924
51	0.0	.13559	.15630	.17468	.19645	.21147
	0.5	.13716	.15816	.17679	.19884	.21405
	1.0	.13944	.16064	.17947	.20177	.21716
52	0.0	.13443	.15495	.17316	.19473	.20961
	0.5	.13596	.15675	.17521	.19706	.21213
	1.0	.13818	.15917	.17782	.19991	.21515
53	0.0	.13330	.15363	.17168	.19305	.20780
	0.5	.13480	.15539	.17367	.19532	.21025
	1.0	.13695	.15775	.17621	.19809	.21319
54	0.0	.13221	.15235	.17024	.19142	.20604
	0.5	.13366	.15406	.17218	.19363	.20842
	1.0	.13576	.15635	.17465	.19632	.21128
55	0.0	.13113	.15110	.16884	.18983	.20432
	0.5	.13255	.15277	.17072	.19198	.20665
	1.0	.13459	.15500	.17313	.19461	.20943
56	0.0	.13009	.14989	.16746	.18828	.20265
	0.5	.13147	.15151	.16931	.19037	.20491
	1.0	.13346	.15368	.17165	.19293	.20762
57	0.0	.12907	.14870	.16613	.18677	.20101
	0.5	.13041	.15028	.16792	.18881	.20322
	1.0	.13235	.15240	.17021	.19130	.20586
58	0.0	.12807	.14754	.16482	.18529	.19942
	0.5	.12939	.14908	.16657	.18728	.20157
	1.0	.13128	.15115	.16880	.18971	.20415
59	0.0	.12710	.14641	.16355	.18385	.19786
	0.5	.12838	.14791	.16526	.18579	.19997
	1.0	.13022	.14993	.16743	.18816	.20247

TABLE **X** Critical values of the δ-corrected one-sample Kolmogorov–Smirnov statistic (*continued*)

		α				
n	**δ**	**0.2**	**0.1**	**0.05**	**0.02**	**0.01**
60	0.0	.12615	.14530	.16230	.18245	.19635
	0.5	.12740	.14677	.16397	.18434	.19840
	1.0	.12920	.14874	.16609	.18665	.20084
61	0.0	.12522	.14422	.16109	.18107	.19486
	0.5	.12644	.14566	.16272	.18292	.19687
	1.0	.12820	.14757	.16478	.18518	.19925
62	0.0	.12431	.14316	.15990	.17973	.19341
	0.5	.12551	.14457	.16149	.18154	.19537
	1.0	.12722	.14644	.16351	18374	.19770
63	0.0	.12342	.14213	.15874	.17841	.19199
	0.5	.12459	.14350	.16029	.18018	.19391
	1.0	.12627	.14533	.16226	.18233	.19618
64	0.0	.12255	.14112	.15760	.17713	.19061
	0.5	.12370	.14246	.15912	.17886	.19248
	1.0	.12533	.14425	.16105	.18096	.19470
65	0.0	.12170	.14013	.15649	.17587	.18925
	0.5	.12282	.14144	.15798	.17756	.19108
	1.0	.12442	.14319	.15986	.17961	.19325
66	0.0	.12087	.13916	.15540	.17464	.18792
	0.5	.12197	.14045	.15686	.17629	.18971
	1.0	.12353	.14215	.15870	.17830	.19184
67	0.0	.12006	.13821	.15433	.17344	.18662
	0.5	.12113	.13847	.15576	.17505	.18837
	1.0	.12266	.14114	.15756	.17702	.19045
68	0.0	.11926	.13728	.15329	.17226	.18535
	0.5	.12031	.13851	.15469	.17384	.18706
	1.0	.12180	.14015	.15645	.17576	.18910
69	0.0	.11848	.13637	.15227	.17110	.18410
	0.5	.11950	.13758	.15363	.17265	.18578
	1.0	.12097	.13918	.15536	.17453	.18777
70	0.0	.11771	.13548	.15127	.16997	.18288
	0.5	.11872	.13666	.15260	.17149	.18452
	1.0	.12015	.13823	.15429	.17333	.18647
71	0.0	.11696	.13461	.15028	.16886	.18168
	0.5	.11794	.13577	.15160	.17035	.18329
	1.0	.11935	.13730	.15325	.17215	.18520
72	0.0	.11622	.13375	.14932	.16777	.18051
	0.5	.11719	.13489	.15061	.16923	.18209
	1.0	.11856	.13639	.15223	.17100	.18396
73	0.0	.11550	.13291	.14838	.16671	.17936
	0.5	.11644	.13402	.14964	.16814	.18091
	1.0	.11779	.13550	.15122	.16987	.18274
74	0.0	.11479	.13208	.14745	.16566	.17823
	0.5	.11572	.13318	.14869	.16706	.17975
	1.0	.11704	.13462	.15024	.16876	.18154
75	0.0	.11409	.13128	.14654	.16463	.17712
	0.5	.11500	.13235	.14775	.16601	.17861
	1.0	.11630	.13376	.14928	.16767	.18037
76	0.0	.11341	.13048	.14565	.16363	.17604
	0.5	.11430	.13153	.14684	.16498	.17750
	1.0	.11557	.13292	.14834	.16661	.17922
77	0.0	.11273	.12970	.14478	.16264	.17497
	0.5	.11361	.13073	.14594	.16396	.17640
	1.0	.11486	.13210	.14741	.16556	.17810
78	0.0	.11208	.12894	.14392	.16167	.17392
	0.5	.11294	.12995	.14506	.16297	.17533
	1.0	.11416	.13129	.14650	.16454	.17699

TABLE **X**   Critical values of the δ-corrected one-sample Kolmogorov–Smirnov statistic (*continued*)

α

n	δ	0.2	0.1	0.05	0.02	0.01
79	0.0	.11143	.12818	.14307	.16071	.17289
	0.5	.11228	.12918	.14420	.16199	.17427
	1.0	.11348	.13049	.14561	.16353	.17591
80	0.0	.11079	.12745	.14224	.15978	.17188
	0.5	.11162	.12842	.14335	.16103	.17324
	1.0	.11280	.12971	.14473	.16254	.17484
81	0.0	.11017	.12672	.14143	.15886	.17089
	0.5	.11098	.12768	.14251	.16009	.17222
	1.0	.11214	.12894	.14387	.16157	.17380
82	0.0	.10955	.12601	.14063	.15795	.16992
	0.5	.11036	.12695	.14169	.15916	.17122
	1.0	.11149	.12819	.14303	.16062	.17277
83	0.0	.10895	.12531	.13984	.15706	.16896
	0.5	.10974	.12623	.14089	.15825	.17024
	1.0	.11086	.12745	.14220	.15969	.17176
84	0.0	.10835	.12462	.13907	.15619	.16802
	0.5	.10913	.12553	.14009	.15736	.16928
	1.0	.11023	.12673	.14139	.15877	.17077
85	0.0	.10777	.12394	.13831	.15533	.16709
	0.5	.10853	.12483	.13932	.15648	.16833
	1.0	.10961	.12601	.14059	.15786	.16980
86	0.0	.10719	.12327	.13756	.15449	.16618
	0.5	.10795	.12415	.13855	.15562	.16740
	1.0	.10901	.12531	.13980	.15698	.16884
87	0.0	.10663	.12262	.13682	.15366	.16528
	0.5	.10737	.12348	.13780	.15477	.16648
	1.0	.10841	.12462	.13903	.15610	.16790
88	0.0	.10607	.12197	.13610	.15284	.16440
	0.5	.10680	.12282	.13706	.15393	.16558
	1.0	.10783	.12394	.13827	.15525	.16697
89	0.0	.10553	.12134	.13538	.15203	.16353
	0.5	.10624	.12217	.13633	.15311	.16470
	1.0	.10725	.12327	.13752	.15440	.16606
90	0.0	.10499	.12071	.13468	.15124	.16268
	0.5	.10569	.12154	.13561	.15230	.16382
	1.0	.10668	.12262	.13678	.15357	.16517
91	0.0	.10446	.12010	.13399	.15046	.16184
	0.5	.10515	.12091	.13491	.15150	.16297
	1.0	.10612	.12197	.13606	.15276	.16429
92	0.0	.10393	.11949	.13331	.14970	.16101
	0.5	.10461	.12029	.13421	.15072	.16212
	1.0	.10558	.12134	.13534	.15195	.16342
93	0.0	.10342	.11889	.13264	.14892	.16020
	0.5	.10409	.11968	.13353	.14995	.16130
	1.0	.10504	.12071	.13464	.15116	.16257
94	0.0	.10291	.11831	.13198	.14820	.15940
	0.5	.10357	.11908	.13286	.14919	.16047
	1.0	.10450	.12010	.13395	.15038	.16173
95	0.0	.10241	.11773	.13133	.14747	.15861
	0.5	.10306	.11849	.13220	.14844	.15966
	1.0	.10398	.11949	.13327	.14962	.16091
96	0.0	.10192	.11716	.13070	.14674	.15783
	0.5	.10256	.11791	.13154	.14771	.15887
	1.0	.10346	.11889	.13260	.14886	.16009
97	0.0	.10144	.11659	.13006	.14603	.15706
	0.5	.10207	.11734	.13090	.14698	.15809
	1.0	.10296	.11830	.13195	.14812	.15929

TABLE X Critical values of the δ-corrected one-sample Kolmogorov–Smirnov statistic (*continued*)

α

n	δ	0.2	0.1	0.05	0.02	0.01
98	0.0	.10096	.11604	.12944	.14533	.15631
	0.5	.10158	.11677	.13027	.14627	.15732
	1.0	.10246	.11773	.13130	.14739	.15850
99	0.0	.10049	.11550	.12883	.14464	.15556
	0.5	.10110	.11621	.12964	.14556	.15656
	1.0	.10196	.11715	.13066	.14667	.15773
100	0.0	.10002	.11496	.12823	.14396	.15483
	0.5	.10063	.11567	.12903	.14487	.15581
	1.0	.10148	.11659	.13003	.14596	.15696
>100	0.5	$\dfrac{1.07298}{\sqrt{n}}-\dfrac{1}{2n}$	$\dfrac{1.22387}{\sqrt{n}}-\dfrac{1}{2n}$	$\dfrac{1.35810}{\sqrt{n}}-\dfrac{1}{2n}$	$\dfrac{1.41421}{\sqrt{n}}-\dfrac{1}{2n}$	$\dfrac{1.62762}{\sqrt{n}}-\dfrac{1}{2n}$

TABLE **Y** Critical values of the δ-corrected one-sample Kolmogorov–Smirnov statistic for estimated parameters

This table furnishes critical values of the δ-corrected Kolmogorov–Smirnov test statistic, $g_{\delta,n}$, for testing goodness of fit of an observed cumulative frequency distribution to an expected distribution with parameters estimated from the sample data. Critical values of $g_{\delta,n}$ are given for three values of δ (0. 0.5, and 1) for every sample size from 3 to 100. The probabilities 0.25, 0.1, 0.05, and 0.01 given at the heads of the columns of the table are the probabilities, α, that a sample value of g_δ equals or exceeds the tabulated value when the null hypothesis is true (sample drawn from a population with the expected distribution). The statistic g_δ is the maximum difference between an observed and an expected cumulative frequency distribution plotted at positions $F_\delta = (i - \delta)/(n - 2\delta + 1)$ for the ith observation.

Note that these critical values are not the same as those for tabulations of the traditional D_{max} Kolmogorov–Smirnov statistic for estimated parameters. Critical values of D_{max} can be obtained using the relationship $D_{max} = g_{0.5,n} + 1/2n$. The new approach using δ = 0 and 1 is from H. J. Khamis (*Nonparametric Stat.* **2:**17–27, 1992). It provides a more powerful test.

For a sample of $n = 17$ items, the 5% critical values of g_0 and g_1 are 0.17343 and 0.18746, respectively. If sample values of either g_0 and g_1 equal or exceed their critical values, then the null hypothesis is rejected at the 5% level. For sample sizes $n > 100$, where the effect of δ is relatively unimportant, the critical value of $g_{0.5}$ can be approximated using the relationship shown in the last line of the table. For example, for $\alpha = 0.05$ this value is $0.89196/\sqrt{n} - 1/2n$.

The critical values of $g_{\delta,n}$ are used to test the deviations of an observed continuous frequency distribution from an expected one (Box 17.3, Section 17.2) and for setting confidence limits to a cumulative frequency distribution. The parameters of the expected distribution are estimated from the observed distribution. Appropriate critical values are provided in Table **X** for the less common case of fitting a normal distribution to a sample using known values of the mean and standard deviation.

This table is based on values provided by H. J. Khamis. A description of the table and the Monte Carlo simulations on which it is based is given in Khamis (*Nonparametric Stat.* **2:**17–27, 1992).

TABLE **Y** Critical values of the δ-corrected one-sample
Kolmogorov–Smirnov statistic for estimated parameters

		α			
n	δ	0.25	0.1	0.05	0.01
3	0.0	.16983	.19923	.20904	.21650
	0.5	.16983	.19923	.20904	.21650
	1.0	.24040	.26408	.27302	.28022
4	0.0	.17381	.19982	.22432	.26129
	0.5	.16810	.21871	.24932	.28629
	1.0	.21645	.26101	.29099	.32796
5	0.0	.17201	.20607	.22734	.26604
	0.5	.17773	.22002	.24518	.29888
	1.0	.20719	.25277	.28893	.34888
6	0.0	.16964	.20274	.22065	.26152
	0.5	.17606	.21371	.23857	.28546
	1.0	.19896	.24446	.27332	.32911
7	0.0	.16539	.19691	.21740	.25750
	0.5	.17042	.20790	.23272	.28110
	1.0	.19100	.23322	.26196	.31944
8	0.0	.16135	.19199	.21066	.24892
	0.5	.16646	.20200	.22377	.26640
	1.0	.18470	.22369	.25031	.29682
9	0.0	.15723	.18634	.20515	.24299
	0.5	.16281	.19610	.21671	.26155
	1.0	.17806	.21603	.23780	.28860
10	0.0	.15345	.18222	.20028	.23881
	0.5	.15770	.19028	.21072	.25457
	1.0	.17135	.20607	.23016	.28064
11	0.0	.14947	.17791	.19647	.22899
	0.5	.15379	.18435	.20621	.24444
	1.0	.16608	.20083	.22296	.26523
12	0.0	.14689	.17444	.19318	.22971
	0.5	.15019	.18065	.20329	.24268
	1.0	.16092	.19413	.21809	.26192
13	0.0	.14329	.16988	.18630	.22151
	0.5	.14677	.17592	.19379	.23145
	1.0	.15698	.18795	.20696	.24900
14	0.0	.14056	.16630	.18376	.21565
	0.5	.14373	.17203	.19057	.22453
	1.0	.15222	.18317	.20311	.24008
15	0.0	.13727	.16220	.17999	.21246
	0.5	.13981	.16731	.18648	.22201
	1.0	.14837	.17760	.19735	.23721
16	0.0	.13535	.16062	.17683	.20594
	0.5	.13777	.16508	.18263	.21481
	1.0	.14487	.17379	.19283	.22719
17	0.0	.13280	.15716	.17343	.20379
	0.5	.13566	.16060	.17906	.21199
	1.0	.14217	.16985	.18746	.22709
18	0.0	.13010	.15353	.16916	.19968
	0.5	.13226	.15717	.17419	.20762
	1.0	.13849	.16492	.18288	.21810
19	0.0	.12702	.15069	.16468	.19362
	0.5	.12912	.15436	.16963	.19958
	1.0	.13443	.16110	.17729	.20807
20	0.0	.12501	.14716	.16260	.18906
	0.5	.12684	.15110	.16720	.19631
	1.0	.13137	.15737	.17442	.20377

TABLE **Y** Critical values of the δ-corrected one-sample Kolmogorov–Smirnov statistic for estimated parameters (*continued*)

		α			
n	δ	**0.25**	**0.1**	**0.05**	**0.01**
21	0.0	.12273	.14506	.15980	.19079
	0.5	.12476	.14795	.16389	.19581
	1.0	.12946	.15416	.16971	.20309
22	0.0	.12095	.14304	.15714	.18472
	0.5	.12250	.14581	.16128	.19129
	1.0	.12674	.15102	.16780	.20078
23	0.0	.11966	.13986	.15433	.18127
	0.5	.12066	.14265	.15840	.18578
	1.0	.12482	.14816	.16360	.19292
24	0.0	.11708	.13798	.15173	.18057
	0.5	.11876	.14084	.15481	.18506
	1.0	.12308	.14532	.16000	.19178
25	0.0	.11456	.13508	.14807	.17480
	0.5	.11589	.13777	.15167	.18017
	1.0	.11968	.14182	.15674	.18623
26	0.0	.11407	.13351	.14651	.17383
	0.5	.11505	.13608	.14894	.17850
	1.0	.11870	.14031	.15420	.18423
27	0.0	.11259	.13175	.14395	.16943
	0.5	.11389	.13384	.14689	.17494
	1.0	.11705	.13785	.15158	.18056
28	0.0	.11079	.13086	.14324	.16780
	0.5	.11195	.13296	.14614	.17150
	1.0	.11510	.13684	.15015	.17642
29	0.0	.10879	.12799	.14081	.16648
	0.5	.11025	.12954	.14341	.17035
	1.0	.11340	.13355	.14736	.17613
30	0.0	.10752	.12654	.13871	.16287
	0.5	.10839	.12823	.14082	.16597
	1.0	.11144	.13221	.14468	.17092
31	0.0	.10627	.12476	.13794	.16189
	0.5	.10719	.12653	.13966	.16573
	1.0	.11007	.12974	.14329	.17114
32	0.0	.10531	.12370	.13549	.15900
	0.5	.10632	.12537	.13766	.16297
	1.0	.10925	.12808	.14066	.16706
33	0.0	.10370	.12166	.13370	.15709
	0.5	.10466	.12332	.13576	.15941
	1.0	.10715	.12623	.13909	.16344
34	0.0	.10300	.12089	.13147	.15529
	0.5	.10364	.12255	.13299	.15792
	1.0	.10610	.12508	.13646	.16304
35	0.0	.10167	.11994	.13090	.15333
	0.5	.10281	.12123	.13269	.15533
	1.0	.10521	.12363	.13610	.15943
36	0.0	.10048	.11766	.12897	.15194
	0.5	.10123	.11944	.13120	.15376
	1.0	.10332	.12199	.13421	.15691
37	0.0	.09976	.11660	.12728	.15047
	0.5	.10045	.11784	.12972	.15170
	1.0	.10267	.12070	.13248	.15483
38	0.0	.09758	.11499	.12647	.14818
	0.5	.09821	.11663	.12799	.15024
	1.0	.10038	.11937	.13070	.15313

TABLE **Y** Critical values of the δ-corrected one-sample
Kolmogorov–Smirnov statistic for estimated parameters
(*continued*)

n	δ	α 0.25	0.1	0.05	0.01
39	0.0	.09707	.11415	.12533	.14625
	0.5	.09798	.11539	.12671	.14792
	1.0	.09986	.11819	.12941	.15088
40	0.0	.09545	.11224	.12324	.14600
	0.5	.09623	.11361	.12497	.14817
	1.0	.09821	.11626	.12758	.15141
41	0.0	.09508	.11172	.12299	.14586
	0.5	.09609	.11323	.12480	.14862
	1.0	.09830	.11587	.12782	.15167
42	0.0	.09428	.11142	.12270	.14357
	0.5	.09556	.11310	.12436	.14605
	1.0	.09760	.11532	.12662	.15041
43	0.0	.09365	.10981	.12099	.14283
	0.5	.09461	.11098	.12254	.14514
	1.0	.09656	.11329	.12508	.14902
44	0.0	.09255	.10862	.11947	.14085
	0.5	.09354	.11052	.12084	.14332
	1.0	.09550	.11242	.12356	.14627
45	0.0	.09192	.10812	.11862	.14205
	0.5	.09306	.10956	.12018	.14500
	1.0	.09500	.11128	.12258	.14736
46	0.0	.09120	.10673	.11756	.13779
	0.5	.09207	.10815	.11931	.13943
	1.0	.09361	.11036	.12225	.14239
47	0.0	.09065	.10638	.11656	.13826
	0.5	.09151	.10754	.11808	.14126
	1.0	.09332	.10976	.11998	.14305
48	0.0	.08947	.10509	.11490	.13467
	0.5	.09027	.10635	.11601	.13651
	1.0	.09184	.10823	.11861	.13901
49	0.0	.08877	.10381	.11390	.13462
	0.5	.08964	.10517	.11529	.13660
	1.0	.09150	.10711	.11711	.13891
50	0.0	.08826	.10372	.11320	.13414
	0.5	.08917	.10490	.11474	.13635
	1.0	.09094	.10678	.11711	.13838
51	0.0	.08749	.10341	.11217	.13255
	0.5	.08808	.10427	.11382	.13428
	1.0	.08985	.10578	.11595	.13691
52	0.0	.08629	.10129	.11135	.13061
	0.5	.08697	.10263	.11243	.13234
	1.0	.08858	.10443	.11467	.13523
53	0.0	.085953	.10076	.11040	.12988
	0.5	.086851	.10177	.11136	.13200
	1.0	.088320	.10387	.11305	.13372
54	0.0	.085517	.10011	.10999	.13020
	0.5	.086288	.10141	.11070	.13188
	1.0	.087741	.10306	.11276	.13357
55	0.0	.084609	.09929	.10871	.12737
	0.5	.085430	.10037	.11023	.12841
	1.0	.087083	.10186	.11198	.13054
56	0.0	.083570	.09770	.10709	.12509
	0.5	.084398	.09878	.10807	.12685
	1.0	.085944	.10057	.11002	.12915

TABLE Y Critical values of the δ-corrected one-sample Kolmogorov–Smirnov statistic for estimated parameters (*continued*)

			α		
n	δ	**0.25**	**0.1**	**0.05**	**0.01**
57	0.0	.083269	.09754	.10717	.12533
	0.5	.083772	.09838	.10792	.12726
	1.0	.085347	.10018	.10974	.12933
58	0.0	.082878	.09687	.10629	.12484
	0.5	.083682	.09787	.10752	.12593
	1.0	.084952	.09951	.10921	.12823
59	0.0	.082390	.09647	.10558	.12456
	0.5	.082871	.09736	.10662	.12579
	1.0	.084347	.09911	.10855	.12769
60	0.0	.081941	.09604	.10511	.12300
	0.5	.082529	.09711	.10606	.12423
	1.0	.083697	.09873	.10750	.12606
61	0.0	.081279	.09501	.10429	.12139
	0.5	.081889	.09598	.10502	.12278
	1.0	.083129	.09758	.10659	.12435
62	0.0	.080532	.09489	.10345	.12142
	0.5	.081249	.09593	.10462	.12298
	1.0	.082505	.09748	.10603	.12459
63	0.0	.079697	.09385	.10340	.12102
	0.5	.080220	.09470	.10411	.12297
	1.0	.081574	.09627	.10562	.12451
64	0.0	.079459	.09294	.10207	.12003
	0.5	.080147	.09394	.10300	.12111
	1.0	.081257	.09502	.10440	.12322
65	0.0	.078859	.09230	.10126	.11972
	0.5	.079649	.09319	.10254	.12093
	1.0	.080751	.09460	.10440	.12287
66	0.0	.078505	.09216	.10115	.11911
	0.5	.078956	.09310	.10206	.12071
	1.0	.080224	.09467	.10357	.12236
67	0.0	.077942	.09104	.09996	.11873
	0.5	.078521	.09189	.10082	.11926
	1.0	.079479	.09337	.10210	.12029
68	0.0	.077531	.09052	.09847	.11700
	0.5	.078185	.09101	.09963	.11811
	1.0	.079210	.09210	.10142	.11947
69	0.0	.077131	.09043	.09954	.11709
	0.5	.077864	.09115	.10032	.11799
	1.0	.078785	.09213	.10154	.11950
70	0.0	.076502	.089510	.09796	.11405
	0.5	.077083	.090300	.09883	.11449
	1.0	.078153	.091440	.09999	.11604
71	0.0	.075980	.088710	.09798	.11528
	0.5	.076601	.089425	.09849	.11617
	1.0	.077492	.090732	.09997	.11806
72	0.0	.075696	.088202	.09644	.11476
	0.5	.076116	.088933	.09745	.11549
	1.0	.077058	.089910	.09867	.11662
73	0.0	.074918	.088265	.09661	.11454
	0.5	.075392	.088789	.09752	.11527
	1.0	.076447	.089767	.09881	.11671
74	0.0	.074608	.087186	.09511	.11325
	0.5	.075054	.087947	.09610	.11429
	1.0	.075942	.089103	.09749	.11592

TABLE **Y** Critical values of the δ-corrected one-sample
Kolmogorov–Smirnov statistic for estimated parameters
(*continued*)

n	δ	α 0.25	0.1	0.05	0.01
75	0.0	.074512	.087171	.09536	.11057
	0.5	.074915	.087857	.09578	.11166
	1.0	.075768	.088942	.09711	.11267
76	0.0	.073926	.086469	.09499	.11186
	0.5	.074405	.086674	.09568	.11246
	1.0	.075203	.088013	.09668	.11392
77	0.0	.073640	.085802	.09415	.11068
	0.5	.074034	.086549	.09504	.11148
	1.0	.074867	.087631	.09602	.11238
78	0.0	.072976	.085808	.09406	.11001
	0.5	.073575	.086544	.09488	.11098
	1.0	.074430	.087247	.09586	.11225
79	0.0	.072505	.084959	.09281	.10859
	0.5	.072923	.085674	.09358	.10970
	1.0	.073856	.086642	.09479	.11102
80	0.0	.072030	.084563	.09259	.10935
	0.5	.072547	.085267	.09340	.11037
	1.0	.073403	.086573	.09451	.11141
81	0.0	.072280	.084076	.09218	.10913
	0.5	.072746	.084893	.09259	.10990
	1.0	.073451	.085592	.09387	.11083
82	0.0	.071878	.083610	.09186	.10706
	0.5	.072291	.084448	.09219	.10800
	1.0	.073033	.085630	.09306	.10963
83	0.0	.071020	.083558	.09098	.10719
	0.5	.071389	.084082	.09173	.10734
	1.0	.072210	.085017	.09263	.10847
84	0.0	.070995	.082864	.09089	.10626
	0.5	.071386	.083503	.09155	.10691
	1.0	.072176	.084423	.09255	.10839
85	0.0	.070347	.082062	.09015	.10580
	0.5	.070585	.082848	.09053	.10688
	1.0	.071378	.083509	.09143	.10843
86	0.0	.069751	.081631	.08972	.10556
	0.5	.070073	.082105	.090250	.10629
	1.0	.070964	.083127	.091351	.10747
87	0.0	.069856	.081532	.089295	.10476
	0.5	.070382	.082030	.089892	.10527
	1.0	.071367	.083150	.090905	.10597
88	0.0	.069342	.081063	.088660	.10473
	0.5	.069884	.081433	.089299	.10542
	1.0	.070511	.082300	.090370	.10616
89	0.0	.069201	.080677	.088210	.10359
	0.5	.069500	.081433	.088787	.10445
	1.0	.070233	.082220	.089692	.10533
90	0.0	.068695	.080214	.087727	.10397
	0.5	.069087	.080624	.088091	.10453
	1.0	.069746	.081583	.089153	.10574
91	0.0	.068227	.079899	.087179	.10146
	0.5	.068683	.080283	.087684	.10209
	1.0	.069340	.080893	.088634	.10325
92	0.0	.067897	.079522	.087111	.10212
	0.5	.068224	.080076	.087582	.10321
	1.0	.068909	.080765	.088635	.10381

TABLE Y Critical values of the δ-corrected one-sample Kolmogorov–Smirnov statistic for estimated parameters (*continued*)

		α			
n	δ	**0.25**	**0.1**	**0.05**	**0.01**
93	0.0	.067706	.078969	.086796	.10175
	0.5	.068051	.079775	.087402	.10262
	1.0	.068671	.080662	.088496	.10350
94	0.0	.067577	.078827	.086131	.10147
	0.5	.068063	.079329	.087007	.10182
	1.0	.068653	.080133	.087780	.10286
95	0.0	.067422	.077937	.085812	.10170
	0.5	.067693	.078230	.086296	.10267
	1.0	.068249	.079322	.087178	.10361
96	0.0	.066703	.078093	.085326	.10033
	0.5	.067068	.078498	.086096	.10082
	1.0	.067566	.079182	.086754	.10241
97	0.0	.066482	.077562	.084648	.09792
	0.5	.066827	.078089	.085168	.09969
	1.0	.067474	.079019	.085821	.10039
98	0.0	.065780	.077667	.084833	.09831
	0.5	.066337	.077875	.085257	.09889
	1.0	.067108	.078592	.085991	.09997
99	0.0	.065925	.076855	.084156	.09801
	0.5	.066235	.077303	.084718	.09883
	1.0	.066793	.077966	.085602	.09957
100	0.0	.065677	.076701	.083829	.09852
	0.5	.066016	.077127	.084196	.09927
	1.0	.066677	.077712	.085066	.09988
>100	0.5	$\dfrac{.71016}{\sqrt{n}} - \dfrac{1}{2n}$	$\dfrac{.82127}{\sqrt{n}} - \dfrac{1}{2n}$	$\dfrac{.89196}{\sqrt{n}} - \dfrac{1}{2n}$	$\dfrac{1.0427}{\sqrt{n}} - \dfrac{1}{2n}$

TABLE Z   Critical values for Page's test

This table furnishes critical values for Page's L-test for ordered alternatives. Critical values are furnished for a, the number of treatments, ranging from 2 to 10 and for b, the number of blocks, ranging from 2 to 24. At each intersection of a given treatment and block number, three numbers are furnished: the 0.05, 0.01, and 0.001 critical values of L, the statistic for Page's test.

We find the critical value of L for 10 treatments and 4 blocks to be 1382 at $\alpha = 0.001$. If an observed value of L is greater than this tabled critical value, we can reject the null hypothesis that the rank sums are unrelated to the tabulated order of the treatments. For critical values beyond these arguments, employ the following expression:

$$L_\alpha = \frac{b(a^3 - a)}{12}\left[\frac{t_{(2\alpha)[\infty]}}{\sqrt{b(a - 1)}} + \frac{3(a + 1)}{a - 1}\right]$$

This table is used for Page's L-test for ordered alternatives (Section 14.12), a nonparametric test for linear relationships of treatment means in a randomized-complete-blocks design.

Values in this table for $a \leq 8$ and $b \leq 12$ and $a = 3$ for $13 \leq b \leq 20$ were taken from E. B. Page (*J. Am. Stat. Assn* **58**:216–230, 1963) with permission of the author and publisher. Other values were computed using the normal approximation given above.

TABLE Z Critical values for Page's test

b	nominal α	3	4	5	6	7	8	9	10
	.05	28	58	103	166	252	362	500	669
2	.01	-	60	106	173	261	376	520	696
	.001	-	-	109	178	269	388	544	726
	.05	41	84	150	244	370	532	736	986
3	.01	42	87	155	252	382	549	761	1019
	.001	-	89	160	260	394	567	790	1056
	.05	54	111	197	321	487	701	970	1301
4	.01	55	114	204	331	501	722	999	1339
	.001	56	117	210	341	516	743	1032	1382
	.05	66	137	244	397	603	869	1204	1614
5	.01	68	141	251	409	620	893	1236	1656
	.001	70	145	259	420	637	917	1273	1704
	.05	79	163	291	474	719	1037	1436	1926
6	.01	81	167	299	486	737	1063	1472	1972
	.001	83	172	307	499	757	1090	1512	2025
	.05	91	189	338	550	835	1204	1668	2238
7	.01	93	193	346	563	855	1232	1706	2288
	.001	96	198	355	577	876	1262	1750	2344
	.05	104	214	384	625	950	1371	1899	2548
8	.01	106	220	393	640	972	1401	1940	2602
	.001	109	225	403	655	994	1433	1987	2662
	.05	116	240	431	701	1065	1537	2130	2859
9	.01	119	246	441	717	1088	1569	2174	2915
	.001	121	252	451	733	1113	1603	2223	2980
	.05	128	266	477	777	1180	1703	2361	3169
10	.01	131	272	487	793	1205	1736	2407	3228
	.001	134	278	499	811	1230	1773	2459	3296
	.05	141	292	523	852	1295	1868	2591	3478
11	.01	144	298	534	869	1321	1905	2639	3541
	.001	147	305	546	888	1348	1943	2694	3612
	.05	153	317	570	928	1410	2035	2821	3787
12	.01	156	324	581	946	1437	2072	2872	3852
	.001	160	331	593	965	1465	2112	2929	3927

TABLE Z Critical values for Page's test (*continued*)

nominal b α		*a* 3	4	5	6	7	8	9	10
	.05	165	343	615	1002	1524	2201	3051	4096
13	.01	169	350	628	1022	1553	2240	3104	4164
	.001	172	358	642	1044	1585	2285	3163	4241
	.05	178	368	661	1078	1639	2366	3281	4405
14	.01	181	376	674	1098	1668	2407	3335	4475
	.001	185	384	689	1121	1702	2453	3397	4556
	.05	190	394	707	1153	1753	2532	3511	4713
15	.01	194	402	721	1174	1784	2574	3567	4786
	.001	197	410	736	1197	1818	2622	3631	4869
	.05	202	419	753	1228	1868	2697	3740	5021
16	.01	206	427	767	1249	1899	2740	3798	5097
	.001	210	436	783	1274	1935	2790	3864	5183
	.05	215	445	799	1303	1982	2862	3969	5330
17	.01	218	453	814	1325	2014	2907	4029	5407
	.001	223	463	830	1350	2051	2958	4098	5496
	.05	227	471	845	1378	2097	3028	4199	5637
18	.01	231	479	860	1401	2129	3073	4260	5717
	.001	235	489	876	1427	2167	3126	4330	5808
	.05	239	496	891	1453	2210	3193	4428	5945
19	.01	243	505	906	1476	2245	3240	4491	6027
	.001	248	515	923	1503	2283	3294	4563	6121
	.05	251	522	937	1528	2325	3358	4657	6253
20	.01	256	531	953	1552	2360	3406	4722	6337
	.001	260	541	970	1579	2399	3461	4796	6433
	.05	263	547	983	1603	2439	3523	4885	6560
21	.01	268	556	999	1628	2475	3572	4952	6647
	.001	273	567	1017	1656	2515	3629	5028	6745
	.05	275	573	1029	1678	2553	3687	5114	6868
22	.01	280	582	1045	1703	2589	3738	5182	6956
	.001	285	593	1063	1732	2631	3796	5260	7057
	.05	288	598	1075	1753	2667	3852	5343	7175
23	.01	292	608	1091	1778	2704	3904	5413	7265
	.001	298	619	1110	1808	2747	3963	5492	7368
	.05	300	624	1121	1828	2781	4017	5571	7482
24	.01	305	633	1138	1854	2819	4070	5643	7574
	.001	310	644	1157	1884	2863	4130	5724	7679

TABLE **AA**   Critical values of the number of runs

This table furnishes the critical number of runs (sequences of one or more like elements preceded and/or followed by unlike elements) in a sequence of $n_1 + n_2$ randomly arranged items, n_1 of one kind, n_2 of another. The values of α across the top of the table are 0.005, 0.01, 0.025, 0.05, 0.95, 0.975, 0.99, and 0.995. Any number of runs that is *equal to or less than* the desired critical value at the left half of any row in the table or is *equal to or greater than* the desired critical value in the right half of any row leads to a rejection of the hypothesis of random arrangement. The table is divided into two parts: the first permits n_1 and n_2 to vary up to 20 and does not require equal sample sizes. The second part is for equal sample sizes ($n_1 = n_2$) from 10 to 100.

To find values significant at 5%, proceed as follows. If in a sequence of 27 items there are 10 items of type A and 17 of type B and 20 runs, enter the first part of Table **AA** at $n_1 = 10$ and $n_2 = 17$. (Always make n_1 the smaller sample size.) The critical values in the table for the percentage points 0.025 and 0.975 are 8 and 19, respectively. Thus, 8 runs (or fewer) would indicate a significant departure from randomness, as would 19 or more runs.

This procedure implies a two-tailed significance test in which the alternative hypothesis is either too few or too many runs. For this reason critical percentage points corresponding to proportion $\alpha/2$ and $1 - \alpha/2$ are chosen when one is prepared to accept a type I error of α. For tests of one-tailed hypotheses use critical percentage points corresponding to α or $1 - \alpha$, depending on the tail of the distribution being tested. Dashes in the left half of a row and asterisks in the right half of a row indicate cases in the first part of the table where no critical values of runs are possible.

The second part of the table is used similarly. For example, for a sample of 100 items, of which 50 are of type A and 50 are of type B, the 0.01 and 0.995 critical values, 37 and 65 runs, respectively, can be found in the row for $n_1 = n_2 = n = 50$. For sample sizes exceeding those listed in this table, employ the normal approximations given in Box 18.2.

This table is used to evaluate tests of randomness by means of runs tests (Section 18.2).

This table has been copied from tables II and III by F. S. Swed and C. Eisenhart (*Ann. Math. Stat.* **14**:66–87, 1943), except that 1 has been added to the values in the right half of each row to make the critical values in this table compatible in formulation with those of other statistical tables. Inconsistencies pointed out by these authors in five values of their table III have been corrected.

TABLE AA   Critical values of the number of runs

For sample sizes up to $n_1 \leq n_2 = 20$

Cumulative probability

n_1	n_2	0.005	0.01	0.025	0.05	0.95	0.975	0.99	0.995
2	2	-	-	-	-	*	*	*	*
	3	-	-	-	-	*	*	*	*
	4	-	-	-	-	*	*	*	*
	5	-	-	-	-	*	*	*	*
	6	-	-	-	-	*	*	*	*
	7	-	-	-	-	*	*	*	*
	8	-	-	-	2	*	*	*	*
	9	-	-	-	2	*	*	*	*
	10	-	-	-	2	*	*	*	*
	11	-	-	-	2	*	*	*	*
	12	-	-	2	2	*	*	*	*
	13	-	-	2	2	*	*	*	*
	14	-	-	2	2	*	*	*	*
	15	-	-	2	2	*	*	*	*
	16	-	-	2	2	*	*	*	*
	17	-	-	2	2	*	*	*	*
	18	-	-	2	2	*	*	*	*
	19	-	2	2	2	*	*	*	*
	20	-	2	2	2	*	*	*	*
3	3	-	-	-	-	*	*	*	*
	4	-	-	-	-	7	*	*	*
	5	-	-	-	2	*	*	*	*
	6	-	-	2	2	*	*	*	*
	7	-	-	2	2	*	*	*	*
	8	-	-	2	2	*	*	*	*
	9	-	2	2	2	*	*	*	*
	10	-	2	2	3	*	*	*	*
	11	-	2	2	3	*	*	*	*
	12	2	2	2	3	*	*	*	*
	13	2	2	2	3	*	*	*	*
	14	2	2	2	3	*	*	*	*
	15	2	2	3	3	*	*	*	*
	16	2	2	3	3	*	*	*	*
	17	2	2	3	3	*	*	*	*
	18	2	2	3	3	*	*	*	*
	19	2	2	3	3	*	*	*	*
	20	2	2	3	3	*	*	*	*

TABLE AA Critical values of the number of runs (*continued*)

For sample sizes up to $n_1 \leq n_2 = 20$ (*continued*)

Cumulative probability

n_1	n_2	0.005	0.01	0.025	0.05	0.95	0.975	0.99	0.995
4	4	-	-	-	2	8	*	*	*
	5	-	-	2	2	9	9	9	*
	6	-	2	2	3	9	9	*	*
	7	-	2	2	3	9	*	*	*
	8	2	2	3	3	*	*	*	*
	9	2	2	3	3	*	*	*	*
	10	2	2	3	3	*	*	*	*
	11	2	2	3	3	*	*	*	*
	12	2	3	3	4	*	*	*	*
	13	2	3	3	4	*	*	*	*
	14	2	3	3	4	*	*	*	*
	15	3	3	3	4	*	*	*	*
	16	3	3	4	4	*	*	*	*
	17	3	3	4	4	*	*	*	*
	18	3	3	4	4	*	*	*	*
	19	3	3	4	4	*	*	*	*
	20	3	3	4	4	*	*	*	*
5	5	-	2	2	3	9	10	10	*
	6	2	2	3	3	10	10	11	11
	7	2	2	3	3	10	11	11	*
	8	2	2	3	3	11	11	*	*
	9	2	3	3	4	11	*	*	*
	10	3	3	3	4	11	*	*	*
	11	3	3	3	4	11	*	*	*
	12	3	3	3	4	11	*	*	*
	13	3	3	4	4	*	*	*	*
	14	3	3	4	5	*	*	*	*
	15	3	4	4	5	*	*	*	*
	16	3	4	4	5	*	*	*	*
	17	3	4	4	5	*	*	*	*
	18	4	4	5	5	*	*	*	*
	19	4	4	5	5	*	*	*	*
	20	4	4	5	5	*	*	*	*

TABLE **AA** Critical values of the number of runs (*continued*)

For sample sizes up to $n_1 \leq n_2 = 20$ (*continued*)

Cumulative probability

n_1	n_2	0.005	0.01	0.025	0.05	0.95	0.975	0.99	0.995
6	6	2	2	3	3	11	11	12	12
	7	2	3	3	4	11	12	12	13
	8	3	3	3	4	12	12	13	13
	9	3	3	4	4	12	13	13	*
	10	3	3	4	5	12	13	*	*
	11	3	4	4	5	13	13	*	*
	12	3	4	4	5	13	13	*	*
	13	3	4	5	5	13	*	*	*
	14	4	4	5	5	13	*	*	*
	15	4	4	5	6	*	*	*	*
	16	4	4	5	6	*	*	*	*
	17	4	5	5	6	*	*	*	*
	18	4	5	5	6	*	*	*	*
	19	4	5	6	6	*	*	*	*
	20	4	5	6	6	*	*	*	*
7	7	3	3	3	4	12	13	13	13
	8	3	3	4	4	13	13	14	14
	9	3	4	4	5	13	14	14	15
	10	3	4	5	5	13	14	15	15
	11	4	4	5	5	14	14	15	15
	12	4	4	5	6	14	14	15	*
	13	4	5	5	6	14	15	*	*
	14	4	5	5	6	14	15	*	*
	15	4	5	6	6	15	15	*	*
	16	5	5	6	6	15	*	*	*
	17	5	5	6	7	15	*	*	*
	18	5	5	6	7	15	*	*	*
	19	5	6	6	7	15	*	*	*
	20	5	6	6	7	15	*	*	*

TABLE AA Critical values of the number of runs (*continued*)

For sample sizes up to $n_1 \leq n_2 = 20$ (*continued*)

Cumulative probability

n_1	n_2	0.005	0.01	0.025	0.05	0.95	0.975	0.99	0.995
8	8	3	4	4	5	13	14	14	15
	9	3	4	5	5	14	14	15	15
	10	4	4	5	6	14	15	15	16
	11	4	5	5	6	15	15	16	16
	12	4	5	6	6	15	16	16	17
	13	5	5	6	6	15	16	17	17
	14	5	5	6	7	16	16	17	17
	15	5	5	6	7	16	16	17	*
	16	5	6	6	7	16	17	17	*
	17	5	6	7	7	16	17	*	*
	18	6	6	7	8	16	17	*	*
	19	6	6	7	8	16	17	*	*
	20	6	6	7	8	17	17	*	*
9	9	4	4	5	6	14	15	16	16
	10	4	5	5	6	15	16	16	17
	11	5	5	6	6	15	16	17	17
	12	5	5	6	7	16	16	17	18
	13	5	6	6	7	16	17	18	18
	14	5	6	7	7	17	17	18	18
	15	6	6	7	8	17	18	18	19
	16	6	6	7	8	17	18	18	19
	17	6	7	7	8	17	18	19	19
	18	6	7	8	8	18	18	19	*
	19	6	7	8	8	18	18	19	*
	20	7	7	8	9	18	18	19	*
10	10	5	5	6	6	16	16	17	17
	11	5	5	6	7	16	17	18	18
	12	5	6	7	7	17	17	18	19
	13	5	6	7	8	17	18	19	19
	14	6	6	7	8	17	18	19	19
	15	6	7	7	8	18	18	19	20
	16	6	7	8	8	18	19	20	20
	17	7	7	8	9	18	19	20	20
	18	7	7	8	9	19	19	20	21
	19	7	8	8	9	19	20	20	21
	20	7	8	9	9	19	20	20	21

TABLE AA Critical values of the number of runs (*continued*)

For sample sizes up to $n_1 \leq n_2 = 20$ (*continued*)

		Cumulative probability							
n_1	n_2	**0.005**	**0.01**	**0.025**	**0.05**	**0.95**	**0.975**	**0.99**	**0.995**
11	11	5	6	7	7	17	17	18	19
	12	6	6	7	8	17	18	19	19
	13	6	6	7	8	18	19	19	20
	14	6	7	8	8	18	19	20	20
	15	7	7	8	9	19	19	20	21
	16	7	7	8	9	19	20	21	21
	17	7	8	9	9	19	20	21	22
	18	7	8	9	10	20	20	21	22
	19	8	8	9	10	20	21	22	22
	20	8	8	9	10	20	21	22	22
12	12	6	7	7	8	18	19	19	20
	13	6	7	8	9	18	19	20	21
	14	7	7	8	9	19	20	21	21
	15	7	8	8	9	19	20	21	22
	16	7	8	9	10	20	21	22	22
	17	8	8	9	10	20	21	22	22
	18	8	8	9	10	21	21	22	23
	19	8	9	10	10	21	22	23	23
	20	8	9	10	11	21	22	23	23
13	13	7	7	8	9	19	20	21	21
	14	7	8	9	9	20	20	21	22
	15	7	8	9	10	20	21	22	22
	16	8	8	9	10	21	21	22	23
	17	8	9	10	10	21	22	23	23
	18	8	9	10	11	21	22	23	24
	19	9	9	10	11	22	23	24	24
	20	9	10	10	11	22	23	24	24
14	14	7	8	9	10	20	21	22	23
	15	8	8	9	10	21	22	23	23
	16	8	9	10	11	21	22	23	24
	17	8	9	10	11	22	23	24	24
	18	9	9	10	11	22	23	24	25
	19	9	10	11	12	23	23	24	25
	20	9	10	11	12	23	24	25	25

TABLE AA Critical values of the number of runs (*continued*)

For sample sizes up to $n_1 \leq n_2 = 20$ (*continued*)

Cumulative probability

n_1	n_2	0.005	0.01	0.025	0.05	0.95	0.975	0.99	0.995
15	15	8	9	10	11	21	22	23	24
	16	9	9	10	11	22	23	24	24
	17	9	10	11	11	22	23	24	25
	18	9	10	11	12	23	24	25	25
	19	10	10	11	12	23	24	25	26
	20	10	11	12	12	24	25	26	26
16	16	9	10	11	11	23	23	24	25
	17	9	10	11	12	23	24	25	26
	18	10	10	11	12	24	25	26	26
	19	10	11	12	13	24	25	26	27
	20	10	11	12	13	25	25	26	27
17	17	10	10	11	12	24	25	26	26
	18	10	11	12	13	24	25	26	27
	19	10	11	12	13	25	26	27	27
	20	11	11	13	13	25	26	27	28
18	18	11	11	12	13	25	26	27	27
	19	11	12	13	14	25	26	27	28
	20	11	12	13	14	26	27	28	29
19	19	11	12	13	14	26	27	28	29
	20	12	12	13	14	27	27	29	29
20	20	12	13	14	15	27	28	29	30

TABLE AA Critical values of the number of runs (*continued*)

For sample sizes up to $n_1 \leq n_2 = 20$ (*continued*)

Cumulative probability

n	0.005	0.01	0.025	0.05	0.95	0.975	0.99	0.995
10	5	5	6	6	16	16	17	17
11	5	6	7	7	17	17	18	19
12	6	6	7	8	18	19	20	20
13	7	7	8	9	19	20	21	21
14	7	8	9	10	20	21	22	23
15	8	9	10	11	21	22	23	24
16	9	10	11	11	23	23	24	25
17	10	10	11	12	24	25	26	26
18	10	11	12	13	25	26	27	28
19	11	12	13	14	26	27	28	29
20	12	13	14	15	27	28	29	30
21	13	14	15	16	28	29	30	31
22	14	14	16	17	29	30	32	32
23	14	15	16	17	31	32	33	34
24	15	16	17	18	32	33	34	35
25	16	17	18	19	33	34	35	36
26	17	18	19	20	34	35	36	37
27	18	19	20	21	35	36	37	38
28	18	19	21	22	36	37	39	40
29	19	20	22	23	37	38	40	41
30	20	21	22	24	38	40	41	42
31	21	22	23	25	39	41	42	43
32	22	23	24	25	41	42	43	44
33	23	24	25	26	42	43	44	45
34	23	24	26	27	43	44	46	47
35	24	25	27	28	44	45	47	48
36	25	26	28	29	45	46	48	49
37	26	27	29	30	46	47	49	50
38	27	28	30	31	47	48	50	51
39	28	29	30	32	48	50	51	52
40	29	30	31	33	49	51	52	53
41	29	31	32	34	50	52	53	55
42	30	31	33	35	51	53	55	56
43	31	32	34	35	53	54	56	57
44	32	33	35	36	54	55	57	58
45	33	34	36	37	55	56	58	59
46	34	35	37	38	56	57	59	60
47	35	36	38	39	57	58	60	61
48	35	37	38	40	58	60	61	63
49	36	38	39	41	59	61	62	64
50	37	38	40	42	60	62	64	65
51	38	39	41	43	61	63	65	66
52	39	40	42	44	62	64	66	67
53	40	41	43	45	63	65	67	68
54	41	42	44	45	65	66	68	69

TABLE AA Critical values of the number of runs (*continued*)

For sample sizes up to $n_1 \le n_2 = 20$ (*continued*)

Cumulative probability

n	0.005	0.01	0.025	0.05	0.95	0.975	0.95	0.995
55	42	43	45	46	66	67	69	70
56	42	44	46	47	67	68	70	72
57	43	45	47	48	68	69	71	73
58	44	46	47	49	69	71	72	74
59	45	46	48	50	70	72	74	75
60	46	47	49	51	71	73	75	76
61	47	48	50	52	72	74	76	77
62	48	49	51	53	73	75	77	78
63	49	50	52	54	74	76	78	79
64	49	51	53	55	75	77	79	81
65	50	52	54	56	76	78	80	82
66	51	53	55	57	77	79	81	83
67	52	54	56	58	78	80	82	84
68	53	54	57	58	80	81	84	85
69	54	55	58	59	81	82	85	86
70	55	56	58	60	82	84	86	87
71	56	57	59	61	83	85	87	88
72	57	58	60	62	84	86	88	89
73	57	59	61	63	85	87	89	91
74	58	60	62	64	86	88	90	92
75	59	61	63	65	87	89	91	93
76	60	62	64	66	88	90	92	94
77	61	63	65	67	89	91	93	95
78	62	64	66	68	90	92	94	96
79	63	64	67	69	91	93	96	97
80	64	65	68	70	92	94	97	98
81	65	66	69	71	93	95	98	99
82	66	67	69	71	95	97	99	100
83	66	68	70	72	96	98	100	102
84	67	69	71	73	97	99	101	103
85	68	70	72	74	98	100	102	104
86	69	71	73	75	99	101	103	105
87	70	72	74	76	100	102	104	106
88	71	73	75	77	101	103	105	107
89	72	74	76	78	102	104	106	108
90	73	74	77	79	103	105	108	109
91	74	75	78	80	104	106	109	110
92	75	76	79	81	105	107	110	111
93	75	77	80	82	106	108	111	113
94	76	78	81	83	107	109	112	114
95	77	79	82	84	108	110	113	115
96	78	80	82	85	109	112	114	116
97	79	81	83	86	110	113	115	117
98	80	82	84	87	111	114	116	118
99	81	83	85	87	113	115	117	119
100	82	84	86	88	114	116	118	120

TABLE **BB**   Critical values for runs up and down

This table furnishes the critical number of runs up and down (sequences of successive differences of like sign) in a sequence of n randomly arranged items. The values of α across the top of the table are 0.01, 0.025, 0.05, 0.95, 0.975, and 0.99. The sample size n ranged from 5 to 25. Any number of runs that is equal to or less than the critical value at the left half of any row in the table or is equal to or greater than the critical value at the right half of any row leads to a rejection of the hypothesis of random arrangement of successive differences with a one-tail type I error rate of α or $1 - \alpha$.

For a two-tailed significance test in which the alternative hypothesis is either too few or too many runs, use percentage points corresponding to the proportions of $\alpha/2$ and $1 - \alpha/2$ when prepared to accept a type I error of α. To find critical values significant at the 5% level, proceed as follows. For a sequence of 20 observations enter the table at $n = 20$ and note the values 8 and 16 in the columns labeled 0.025 and 0.975, respectively. Thus, as few as 8 runs or as many as 16 runs up and down indicate a significant departure from randomness at the 5% level in a sample of 20 items. For tests of one-tailed hypotheses use the percentage points corresponding to α or $1 - \alpha$, depending on the tail of the distribution being tested. Dashes in the left half of a row and asterisks in the right half of a row indicate cases in the upper part of the table where no significant numbers of runs are possible. For sample sizes $n > 25$, employ the asymptotic approximation given in Box 18.3.

This table is used to evaluate the randomness of trends by the test for runs up and down (Section 18.2).

The table was constructed from information furnished in the table X of J. V. Bradley, *Distribution-Free Statistical Tests* (Prentice-Hall, New Jersey, 1968).

TABLE BB Critical values for runs up and down

Cumulative probability

n	0.01	0.025	0.05	0.95	0.975	0.99
5	-	1	1	*	*	*
6	1	1	1	*	*	*
7	1	2	2	*	*	*
8	2	2	2	*	*	*
9	2	3	3	7	*	*
10	3	3	3	8	*	*
11	3	4	4	9	9	*
12	4	4	4	10	10	*
13	4	5	5	11	11	11
14	5	6	6	11	12	12
15	5	6	6	12	13	13
16	6	7	7	13	13	14
17	6	7	7	14	14	15
18	7	7	8	14	15	15
19	7	8	8	15	16	16
20	8	8	9	16	16	17
21	8	9	10	17	17	18
22	9	10	10	17	18	19
23	10	10	11	18	18	19
24	10	11	11	19	19	20
25	11	11	12	20	20	21

TABLE CC   Critical values for testing outliers (according to Dixon)

This table features critical values for Dixon's test statistic for outliers. Critical values are furnished for sample sizes from 3 to 25 and for α values of 0.10, 0.05, and 0.01. The table is divided into four parts for four different test statistics, shown in the right-hand column. The probabilities given are one-tailed. For a two-tailed test use the same critical values but double the probability at the heads of the columns.

To find the 5% critical value for a sample of 12, employ test statistic r_{21} where Y_1 is the suspected outlier and the other subscripted values of Y are the respective variates in an ordered array of the variates, the first of which is the suspected outlier. Thus, the 5% critical value of r_{21} is 0.546. If the observed test statistic equals or exceeds that value, we consider it an outlier.

This table is used when employing Dixon's test for outliers in normal populations (Section 13.4).

The table was extracted from a more extensive one (table A-8e) in W. J. Dixon and F. J. Massey, Jr., *Introduction to Statistical Analysis,* 3d ed. (McGraw-Hill, New York, 1969) with permission of the publisher.

TABLE CC Critical values for testing outliers (according to Dixon)

n	0.10	0.05	0.01	Test statistic
3	0.886	0.941	0.988	
4	0.679	0.765	0.889	
5	0.557	0.642	0.780	$r_{10} = \dfrac{Y_2 - Y_1}{Y_n - Y_1}$
6	0.482	0.560	0.698	
7	0.434	0.507	0.637	
8	0.479	0.554	0.683	
9	0.441	0.512	0.635	$r_{11} = \dfrac{Y_2 - Y_1}{Y_{n-1} - Y_1}$
10	0.409	0.477	0.597	
11	0.517	0.576	0.679	
12	0.490	0.546	0.642	$r_{21} = \dfrac{Y_3 - Y_1}{Y_{n-1} - Y_1}$
13	0.467	0.521	0.615	
14	0.492	0.546	0.641	
15	0.472	0.525	0.616	
16	0.454	0.507	0.595	
17	0.438	0.490	0.577	
18	0.424	0.475	0.561	
19	0.412	0.462	0.547	$r_{22} = \dfrac{Y_3 - Y_1}{Y_{n-2} - Y_1}$
20	0.401	0.450	0.535	
21	0.391	0.440	0.524	
22	0.382	0.430	0.514	
23	0.374	0.421	0.505	
24	0.367	0.413	0.497	
25	0.360	0.406	0.489	

α

TABLE **DD** Critical values for testing outliers (according to Grubbs)

This table features critical values for Grubbs's test statistic for outliers. Critical values are tabled for sample sizes $n = 3$ to $n = 40$ in increments of one, and for sample sizes $n = 40$ to $n = 140$ in increments of 10. The one-tailed probabilities furnished are 0.10, 0.05, 0.025, 0.01, and 0.005. For a two-tailed test use the same critical values but double the probabilities α.

Grubbs's test statistic is $(Y_1 - \overline{Y})/s$, where Y_1 is the suspected outlier, $\overline{Y}$ is the sample mean, and s is the sample standard deviation. Whenever the outlier is below the mean, consider the ratio as positive before comparing it with critical values in the table. For a sample of 37 items, then, we find a 5% critical value of 2.835.

This table is useful when carrying out Grubbs's test for outliers in normal samples (Section 13.4).

The table was extracted from a more extensive one (table I) in F. E. Grubbs and G. Beck (*Technometrics* **14**:847–854, 1972) with permission of the authors and publisher.

TABLE DD   Critical values for testing outliers (according to Grubbs)

n	α 0.10	0.05	0.025	0.01	0.005
3	1.148	1.153	1.155	1.155	1.155
4	1.425	1.463	1.481	1.492	1.496
5	1.602	1.672	1.715	1.749	1.764
6	1.729	1.822	1.887	1.944	1.973
7	1.828	1.938	2.020	2.097	2.139
8	1.909	2.032	2.126	2.221	2.274
9	1.977	2.110	2.215	2.323	2.387
10	2.036	2.176	2.290	2.410	2.482
11	2.088	2.234	2.355	2.485	2.564
12	2.134	2.285	2.412	2.550	2.636
13	2.175	2.331	2.462	2.607	2.699
14	2.213	2.371	2.507	2.659	2.755
15	2.247	2.409	2.549	2.705	2.806
16	2.279	2.443	2.585	2.747	2.852
17	2.309	2.475	2.620	2.785	2.894
18	2.335	2.504	2.651	2.821	2.932
19	2.361	2.532	2.681	2.854	2.968
20	2.385	2.557	2.709	2.884	3.001
21	2.408	2.580	2.733	2.912	3.031
22	2.429	2.603	2.758	2.939	3.060
23	2.448	2.624	2.781	2.963	3.087
24	2.467	2.644	2.802	2.987	3.112
25	2.486	2.663	2.822	3.009	3.135
26	2.502	2.681	2.841	3.029	3.157
27	2.519	2.698	2.859	3.049	3.178
28	2.534	2.714	2.876	3.068	3.199
29	2.549	2.730	2.893	3.085	3.218
30	2.563	2.745	2.908	3.103	3.236
31	2.577	2.759	2.924	3.119	3.253
32	2.591	2.773	2.938	3.135	3.270
33	2.604	2.786	2.952	3.150	3.286
34	2.616	2.799	2.965	3.164	3.301
35	2.628	2.811	2.979	3.178	3.316
36	2.639	2.823	2.991	3.191	3.330
37	2.650	2.835	3.003	3.204	3.343
38	2.661	2.846	3.014	3.216	3.356
39	2.671	2.857	3.025	3.228	3.369
40	2.682	2.866	3.036	3.240	3.381
50	2.768	2.956	3.128	3.336	3.483
60	2.837	3.025	3.199	3.411	3.560
70	2.893	3.082	3.257	3.471	3.622
80	2.940	3.130	3.305	3.521	3.673
90	2.981	3.171	3.347	3.563	3.716
100	3.017	3.207	3.383	3.600	3.754
110	3.049	3.239	3.415	3.632	3.787
120	3.078	3.267	3.444	3.662	3.817
130	3.104	3.294	3.470	3.688	3.843
140	3.129	3.318	3.493	3.712	3.867

TABLE EE   Orthogonal polynomials

This table furnishes coefficients of first-, second-, and third-order orthogonal polynomials for sample sizes from $n = 3$ to $n = 12$. These orthogonal polynomials are numerical values of the following polynomial expressions $\xi_1' = \lambda_1 x$; $\xi_2' = \lambda_2[x^2 - \frac{1}{12}(n^2 - 1)]$; and $\xi_3' = \lambda_3[x^3 - \frac{1}{20}(3n^2 - 7)x]$, where the x's are deviations $x = X - \bar{X}$ for the integral values of $X = 1, 2, 3, \ldots, n$ and the λ_i's are arbitrary coefficients multiplying the polynomials to make them into simple integers. The table gives the appropriate numerical values for the n coefficients of the polynomials ξ_1', ξ_2', and ξ_3' for sample sizes from $n = 3$ to $n = 12$. Below each column of coefficients is the quantity $\Sigma\xi_i'^2$, and underneath it the numerical value of coefficient λ_i.

To look up a series of polynomials in this table, select the columns corresponding to the number of (equally spaced) X-points in the sample. For example, for $n = 8$ equally spaced points, the coefficients of the linear (first-order) orthogonal polynomials will be -7, -5, -3, -1, $+1$, $+3$, $+5$, $+7$ for the eight X-values arrayed by increasing order of magnitude. Similarly, the coefficients of the second-degree polynomials will be $+7$, $+1$, -3, -5, -5, -3, $+1$, $+7$, and the third-order polynomials will be -7, $+5$, $+7$, $+3$, -3, -7, -5, $+7$. The corresponding quantities $\Sigma\xi_i'^2$ are 168, 168, and 264, and the coefficients λ_i are 2, 1, and $\frac{2}{3}$, respectively.

The orthogonal polynomials tabled here can be employed for making linear, quadratic, and cubic contrasts among treatment effects in an analysis of variance (Section 14.10), for estimating quadratic and cubic components of regression (Section 16.6), and for fitting quadratic and cubic regression equations by the method of orthogonal polynomials.

This table was generated by means of Tschebyscheff polynomials for discrete variables for the case of equal weights. This method yields a least squares fit for the n points, where n equals the sample size. We obtain the following polynomial expressions for $X = 0, 1, 2, \ldots, n - 1$:

$$F_0(X) = 1 \qquad F_1(X) = 1 - \frac{2X}{n - 1}$$

$$F_{k+1}(X) = \frac{[(2k + 1)(n - 2X - 1)F_k(X)] - [k(n + k)F_{k-1}(X)]}{(k + 1)(n - k - 1)}$$

This expression leads to a recurrence relationship in which the fractions are simplified in the interest of presenting a simpler table. More extensive tables of orthogonal polynomials furnishing polynomials up to the sixth degree and for sample sizes up to $n = 45$ can be found in table XXIII of R. A. Fisher and F. Yates, *Statistical Tables for Biological, Agricultural and Medical Research,* 5th ed. (Oliver & Boyd, Edinburgh, 1958).

TABLE EE   Orthogonal polynomials

n = 3

ξ_1'	ξ_2'
−1	+1
0	−2
+1	+1

$\Sigma\xi_i'^2$ 2   6

λ_i 1   3

n = 4

ξ_1'	ξ_2'	ξ_3'
−3	+1	−1
−1	−1	+3
+1	−1	−3
+3	+1	+1

$\Sigma\xi_i'^2$ 20   4   20

λ_i 2   1   10/3

n = 5

ξ_1'	ξ_2'	ξ_3'
−2	+2	−1
−1	−1	+2
0	−2	0
+1	−1	−2
+2	+2	+1

$\Sigma\xi_i'^2$ 10   14   10

λ_i 1   1   5/6

n = 6

ξ_1'	ξ_2'	ξ_3'
−5	+5	−5
−3	−1	+7
−1	−4	+4
+1	−4	−4
+3	−1	−7
+5	+5	+5

$\Sigma\xi_i'^2$ 70   84   180

λ_i 2   3/2   5/3

n = 7

ξ_1'	ξ_2'	ξ_3'
−3	+5	−1
−2	0	+1
−1	−3	+1
0	−4	0
+1	−3	−1
+2	0	−1
+3	+5	+1

$\Sigma\xi_i'^2$ 28   84   6

λ_i 1   1   1/6

n = 8

ξ_1'	ξ_2'	ξ_3'
−7	+7	−7
−5	+1	+5
−3	−3	+7
−1	−5	+3
+1	−5	−3
+3	−3	−7
+5	+1	−5
+7	+7	+7

$\Sigma\xi_i'^2$ 168   168   264

λ_i 2   1   2/3

n = 9

ξ_1'	ξ_2'	ξ_3'
−4	+28	−14
−3	+ 7	+ 7
−2	− 8	+13
−1	−17	+ 9
0	−20	0
+1	−17	− 9
+2	− 8	−13
+3	+ 7	− 7
+4	+28	+14

$\Sigma\xi_i'^2$ 60   2772   990

λ_i 1   3   5/6

n = 10

ξ_1'	ξ_2'	ξ_3'
−9	+6	−42
−7	+2	+14
−5	−1	+35
−3	−3	+31
−1	−4	+12
+1	−4	−12
+3	−3	−31
+5	−1	−35
+7	+2	−14
+9	+6	+42

$\Sigma\xi_i'^2$ 330   132   8580

λ_i 2   1/2   5/3

n = 11

ξ_1'	ξ_2'	ξ_3'
−5	+15	−30
−4	+ 6	+ 6
−3	− 1	+22
−2	− 6	+23
−1	− 9	+14
0	−10	0
+1	− 9	−14
+2	− 6	−23
+3	− 1	−22
+4	+ 6	− 6
+5	+15	+30

$\Sigma\xi_i'^2$ 110   858   4290

λ_i 1   1   5/6

n = 12

ξ_1'	ξ_2'	ξ_3'
−11	+55	−33
− 9	+25	+ 3
− 7	+ 1	+21
− 5	−17	+25
− 3	−29	+19
− 1	−35	+ 7
+ 1	−35	− 7
+ 3	−29	−19
+ 5	−17	−25
+ 7	+ 1	−21
+ 9	+25	− 3
+11	+55	+35

$\Sigma\xi_i'^2$ 572   12012   5140

λ_i 2   3   2/3

TABLE **FF**   Ten thousand random digits

The 10,000 random digits are arranged in 10 columns of 5 digits and blocked out in rows of 5, providing 25 digits per block.

The table of random digits should be entered at random. Choose the page by a random procedure, determine the row and column number by blindly pointing to it with a pencil, and proceed from there in some predetermined fashion, either horizontally or vertically.

A table of random numbers is useful for a variety of sampling operations. Extended use of the same set of random numbers in the same sampling experiment is ill advised. In such cases, new random numbers should be looked up in a different table or preferably generated on the computer.

A sequence of pseudorandom numbers between 0 and 1 was generated by the multiplicative congruential method. Since a computer with a 36-bit word length was used, the following recurrence equation was employed:

$$x_{i+1} = [x_i \times 5^{13}] \bmod (2^{35})$$

where x_i is the ith pseudorandom number using integer arithmetic. Mod (2^{35}) means divide $x_i \times 5^{13}$ by 2^{35} and use the remainder, not the quotient, in subsequent computations. These numbers were then converted to floating point and divided by 2^{35} to scale them to between 0 and 1. Next, they were multiplied by 10 and truncated for the table. Since an analysis of the table revealed that some digits (3 and 7) were slightly too common and the digits 2 and 8 were a little too rare (although within the range of ordinary sampling error), a transformation was developed to adjust the sequence of numbers so that they would be more uniformly distributed.

TABLE FF Ten thousand random digits

	1	2	3	4	5	6	7	8	9	10	
1	48461	14952	72619	73689	52059	37086	60050	86192	67049	64739	1
2	76534	38149	49692	31366	52093	15422	20498	33901	10319	43397	2
3	70437	25861	38504	14752	23757	59660	67844	78815	23758	86814	3
4	59584	03370	42806	11393	71722	93804	09095	07856	55589	46020	4
5	04285	58554	16085	51555	27501	73883	33427	33343	45507	50063	5
6	77340	10412	69189	85171	29082	44785	83638	02583	96483	76553	6
7	59183	62687	91778	80354	23512	97219	65921	02035	59847	91403	7
8	91800	04281	39979	03927	82564	28777	59049	97532	54540	79472	8
9	12066	24817	81099	48940	69554	55925	48379	12866	51232	21580	9
10	69907	91751	53512	23748	65906	91385	84983	27915	48491	91068	10
11	80467	04873	54053	25955	48518	13815	37707	68687	15570	08890	11
12	78057	67835	28302	45048	56761	97725	58438	91528	24645	18544	12
13	05648	39387	78191	88415	60269	94880	58812	42931	71898	61534	13
14	22304	39246	01350	99451	61862	78688	30339	60222	74052	25740	14
15	61346	50269	67005	40442	33100	16742	61640	21046	31909	72641	15
16	66793	37696	27965	30459	91011	51426	31006	77468	61029	57108	16
17	86411	48809	36698	42453	83061	43769	39948	87031	30767	13953	17
18	62098	12825	81744	28882	27369	88183	65846	92545	09065	22655	18
19	68775	06261	54265	16203	23340	84750	16317	88686	86842	00879	19
20	52679	19595	13687	74872	89181	01939	18447	10787	76246	80072	20
21	84096	87152	20719	25215	04349	54434	72344	93008	83282	31670	21
22	63964	55937	21417	49944	38356	98404	14850	17994	17161	98981	22
23	31191	75131	72386	11689	95727	05414	88727	45583	22568	77700	23
24	30545	68523	29850	67833	05622	89975	79042	27142	99257	32349	24
25	52573	91001	52315	26430	54175	30122	31796	98842	37600	26025	25
26	16586	81842	01076	99414	31574	94719	34656	80018	86988	79234	26
27	81841	88481	61191	25013	30272	23388	22463	65774	10029	58376	27
28	43563	66829	72838	08074	57080	15446	11034	98143	74989	26885	28
29	19945	84193	57581	77252	85604	45412	43556	27518	90572	00563	29
30	79374	23796	16919	99691	80276	32818	62953	78831	54395	30705	30
31	48503	26615	43980	09810	38289	66679	73799	48418	12647	40044	31
32	32049	65541	37937	41105	70106	89706	40829	40789	59547	00783	32
33	18547	71562	95493	34112	76895	46766	96395	31718	48302	45893	33
34	03180	96742	61486	43305	34183	99605	67803	13491	09243	29557	34
35	94822	24738	67749	83748	59799	25210	31093	62925	72061	69991	35
36	34330	60599	85828	19152	68499	27977	35611	96240	62747	89529	36
37	43770	81537	59527	95674	76692	86420	69930	10020	72881	12532	37
38	56908	77192	50623	41215	14311	42834	80651	93750	59957	31211	38
39	32787	07189	80539	75927	75475	73965	11796	72140	48944	74156	39
40	52441	78392	11733	57703	29133	71164	55355	31006	25526	55790	40
41	22377	54723	18227	28449	04570	18882	00023	67101	06895	08915	41
42	18376	73460	88841	39602	34049	20589	05701	08249	74213	25220	42
43	53201	28610	87957	21497	64729	64983	71551	99016	87903	63875	43
44	34919	78901	59710	27396	02593	05665	11964	44134	00273	76358	44
45	33617	92159	21971	16901	57383	34262	41744	60891	57624	06962	45
46	70010	40964	98780	72418	52571	18415	64362	90636	38034	04909	46
47	19282	68447	35665	31530	59832	49181	21914	65742	89815	39231	47
48	91429	73328	13266	54898	68795	40948	80808	63887	89939	47938	48
49	97637	78393	33021	05867	86520	45363	43066	00988	64040	09803	49
50	95150	07625	05255	83254	93943	52325	93230	62668	79529	65964	50

TABLE **FF**   Ten thousand random digits (*continued*)

	1	2	3	4	5	6	7	8	9	10	
51	58237	81333	12573	36181	84900	39614	61303	05086	97670	07961	51
52	54789	75554	36795	42649	02971	97584	38223	52643	25027	56849	52
53	55373	14272	62729	25659	84359	02654	08409	52703	88803	31919	53
54	74251	66100	10773	71393	80972	45092	07932	83065	06585	16454	54
55	86077	90904	14779	75116	50267	52217	08539	08345	52750	22815	55
56	13506	84170	08716	28894	20133	99489	87768	55582	96081	20774	56
57	13226	41411	16074	15438	68840	17064	96917	25404	47708	17861	57
58	01642	25456	69804	29277	99473	07912	90488	73325	88266	18082	58
59	23715	70933	37381	20388	34929	96585	14146	81617	36664	25060	59
60	98436	61100	45346	94664	30677	18677	99524	70767	39525	34023	60
61	96571	81879	50387	77316	18874	00763	99457	70858	79674	95618	61
62	70677	59632	22985	95166	54904	61995	63423	65335	13807	96638	62
63	33725	31717	04704	13669	91697	00107	33667	24770	60044	49107	63
64	40910	75631	56653	42858	85768	21254	01295	21507	33687	02404	64
65	67947	27522	14066	14943	19696	93933	52432	90569	14856	30580	65
66	41797	38840	68744	59348	05120	30184	35212	14348	37661	51451	66
67	41770	94218	52578	36238	40575	16793	77152	23382	11570	87276	67
68	34918	50080	97862	84932	57596	33749	78745	73377	72328	63074	68
69	47898	91359	10606	33735	46812	96239	23815	36757	17882	96143	69
70	45021	03882	94463	96369	56001	16348	89408	84563	66422	62636	70
71	92752	67479	72696	20645	78439	11224	17405	27884	15573	12490	71
72	13229	29631	01944	76916	30063	98507	11345	29576	20215	53972	72
73	39111	13161	85208	70273	24016	04960	46728	60292	24831	06403	73
74	50972	92325	72258	57577	32886	56062	62370	99461	64680	38080	74
75	08190	28700	87859	03684	83762	03810	12325	75445	41946	36420	75
76	98140	55201	89156	99277	78211	78692	96992	80163	67882	36674	76
77	14110	02500	54140	43371	06930	26853	56025	73530	97542	19287	77
78	35106	03726	35458	69204	47084	22676	62125	66443	73712	82879	78
79	04446	93672	24118	40460	79678	51259	37345	49666	08518	04251	79
80	42890	46488	42713	76138	82275	59529	98821	60243	51840	02294	80
81	64856	84896	77627	86920	59181	24162	34918	77203	55518	17174	81
82	29739	02885	14169	81125	59048	59396	35494	25220	91424	94750	82
83	70750	97663	18316	16741	75016	01404	78331	76908	68195	18714	83
84	28695	56066	21108	23021	31950	37840	33674	81877	92490	35488	84
85	58892	85844	04181	58470	13348	64277	43838	50362	08531	83388	85
86	79508	78596	96537	20553	41148	37805	14553	09919	11490	70231	86
87	13617	66975	68598	95450	06285	84134	62474	22329	82134	08283	87
88	15123	55351	28631	77941	90178	35876	27833	92494	88899	41558	88
89	34604	18686	05179	31756	47258	14945	98839	82051	86608	07022	89
90	51863	00432	27846	54577	84476	06652	88250	40187	29735	32621	90
91	65489	54535	60256	91285	81743	48426	58351	89166	52478	23935	91
92	86119	57900	04979	16358	06281	25469	02454	20658	08869	29293	92
93	01963	62421	86788	54260	61287	69893	56446	11608	04760	85532	93
94	52164	39397	60568	88382	01561	78861	02849	31033	76875	08260	94
95	91444	35684	80387	44827	71027	16850	04079	02394	49058	60509	95
96	09881	00351	14759	58624	50470	03348	23703	10959	39733	84954	96
97	09045	29309	17864	27687	21691	59354	56599	99735	57626	55645	97
98	13036	51186	32490	45564	53813	25529	66293	50352	41356	44697	98
99	96114	75971	70563	31203	85747	86469	78133	40310	48223	85933	99
100	47657	02006	33820	89370	60192	08248	10221	27754	31323	59019	100

TABLE FF   Ten thousand random digits (*continued*)

	1	2	3	4	5	6	7	8	9	10	
101	20992	99405	64724	68632	51100	05310	71468	25066	51115	11450	101
102	55109	29812	28047	77266	72131	31420	69336	35633	91527	58904	102
103	78521	74623	05802	80834	59282	21784	63995	71057	43191	39005	103
104	04663	28601	24573	98343	72061	11632	11364	50351	97030	03019	104
105	06008	67645	55979	21012	46444	88351	77264	83331	54233	72933	105
106	01971	82763	53029	54861	78099	93127	64114	08341	60164	50548	106
107	69487	17195	47802	76193	54244	75523	50502	23090	64761	32106	107
108	20480	62789	23756	26759	05657	40569	80333	10691	65575	75659	108
109	53736	32574	76781	99231	99766	36951	44813	21302	42681	20511	109
110	30683	57894	57235	19019	86941	22906	62998	31313	68304	99427	110
111	56503	68759	64988	21189	89361	01093	13458	94501	36804	75758	111
112	71015	58374	55700	17283	02822	46847	19644	56298	28363	88692	112
113	01525	06462	32199	28145	69252	56232	35744	26994	71289	25906	113
114	04748	76563	00781	89201	99705	69183	36777	66719	48022	65126	114
115	36337	76371	56419	57462	91591	85204	46652	53152	23325	27949	115
116	50516	33877	60288	52452	32220	99492	38378	77703	62410	53958	116
117	53278	71860	35327	79118	50330	72410	52210	62145	80132	79276	117
118	24396	73460	21715	15875	76225	01362	29941	96873	21765	51302	118
119	90526	85125	87761	67620	87458	53789	86249	09071	91432	93498	119
120	62016	11174	81655	95547	68586	69706	93755	08894	94045	68308	120
121	84388	10946	96056	64407	16812	86694	45320	14494	71454	33194	121
122	79364	91943	12421	13446	22397	96003	82447	57140	66739	15779	122
123	77008	24942	76020	83095	44461	47443	54642	66043	33403	37242	123
124	81770	21470	81822	91417	16685	41100	38863	69628	80463	62659	124
125	30650	00510	39786	37119	51869	57706	80670	69219	60031	57862	125
126	46808	70639	64709	79976	89899	05189	35484	86220	27698	50382	126
127	75690	09668	49236	26127	72363	38179	06614	73630	75445	50183	127
128	66841	26725	81491	47455	60061	24162	67617	61211	08660	62754	128
129	62679	76672	92854	65564	95428	61608	25213	17120	52691	10703	129
130	35996	76325	67090	92344	96575	06735	12066	09040	23352	95665	130
131	58215	30554	42593	51847	04285	91203	60282	48263	03458	92510	131
132	53846	20911	01113	41215	86563	64485	14213	99552	49607	34128	132
133	67847	22692	98162	28309	86439	29258	05728	13132	40198	54449	133
134	27984	58063	39553	84802	65095	95345	24821	39723	12232	56641	134
135	94370	02410	32366	87038	87873	13448	82265	41010	37682	64202	135
136	98352	83237	78099	55733	92336	51756	85103	57257	49016	68938	136
137	38781	77533	50745	21619	83329	64114	24668	50826	50996	35402	137
138	30910	59371	78549	00134	75070	35640	33679	62954	76462	14207	138
139	93383	68965	97175	27810	63231	38153	22138	83818	14553	00792	139
140	44771	50543	46870	84958	59666	78397	67347	45037	30278	34879	140
141	44265	18136	76765	26301	42887	58467	19346	86266	75905	83908	141
142	30942	69742	71236	82089	65346	31553	93314	69647	71233	93093	142
143	12466	26435	22634	69115	06966	12864	84535	80650	59399	80201	143
144	22246	36534	37474	97571	80532	89892	29782	55803	36494	82152	144
145	79721	95215	36459	06287	83524	06356	61171	76170	44472	91543	145
146	60028	91431	93528	17580	60218	61498	86851	05174	08201	45308	146
147	40591	53078	62739	01053	00466	30604	29595	12719	80770	64790	147
148	35984	21080	79400	27098	62417	45236	95779	14806	42222	64628	148
149	24335	70074	60121	78385	69437	23084	42324	04811	36003	35277	149
150	67517	46090	92338	98851	52758	07435	21735	99397	20332	59809	150

TABLE FF   Ten thousand random digits (*continued*)

	1	2	3	4	5	6	7	8	9	10	
151	13961	73549	50445	02797	77988	32480	99116	19559	92686	96188	151
152	40986	23282	40312	03108	08635	96873	37382	67014	99964	13769	152
153	63483	89965	29965	94242	02160	59656	04459	76675	90391	77608	153
154	58874	06054	93183	37271	41717	93385	50687	95813	10015	59534	154
155	22202	93835	75224	56232	29272	50386	50915	32944	84385	48293	155
156	45242	02824	49627	76875	39915	94590	04095	91482	63116	13103	156
157	36815	62809	45636	22306	70080	03830	84709	50946	28711	14240	157
158	11251	93858	66744	09068	08923	46314	16104	66114	52838	60822	158
159	03146	69101	82270	85732	18453	38142	61825	61458	59184	29726	159
160	93303	30006	37277	23975	73283	24325	04092	47153	90553	07140	160
161	06668	91453	74726	75716	16870	65097	12035	55470	57251	65855	161
162	85684	17562	53211	50234	96530	14736	89674	11447	76989	42206	162
163	78948	97463	09494	09356	61285	72337	72064	73890	21476	17576	163
164	96168	90422	35391	89622	81593	63972	37211	43715	11737	34307	164
165	37451	36052	52866	42907	92003	85818	19516	79923	08078	90027	165
166	26390	45437	02061	30574	07526	18571	66706	13281	28401	11476	166
167	43096	57671	66165	01905	36376	33569	29410	16458	14039	97225	167
168	46451	11389	22938	29517	71409	95998	62650	74830	71919	91099	168
169	24125	56973	87209	21228	27479	72709	83161	88214	98627	05266	169
170	79507	12940	01519	07579	78954	31373	11723	22783	98193	05903	170
171	98697	58698	74367	06975	25894	60973	71050	57472	15828	80313	171
172	76404	29382	20189	11287	27324	39284	70368	11591	66657	56035	172
173	43651	96897	98590	39853	90614	92444	67026	43201	01087	47095	173
174	92428	01267	47588	70879	70394	06709	49391	54079	37182	71208	174
175	48368	43909	13164	40940	63680	91273	00550	23383	10394	89394	175
176	03846	41039	99844	62562	38128	69608	06373	81183	53302	05747	176
177	35346	23964	07646	07075	55749	95942	00912	99482	18441	60600	177
178	83186	51111	97339	80202	08062	81871	87530	73461	91760	56926	178
179	58078	94427	95357	45447	69999	25025	61585	89954	38893	69334	179
180	22762	27816	74381	06738	23387	08183	96318	46138	51234	37122	180
181	65354	19914	33459	72776	34215	18749	24371	88366	38555	08983	181
182	79286	15905	28669	50696	19442	26417	60906	05173	34355	97089	182
183	35860	07775	77100	45226	76099	37087	53639	01933	46743	47394	183
184	41998	67048	53397	38732	16130	33833	87778	84418	56556	35785	184
185	08826	99500	16607	64822	13390	55106	37720	56819	46828	10690	185
186	47255	75479	72807	79522	14904	47046	33009	95298	93865	02278	186
187	35684	31325	38516	20620	42438	90223	23470	41134	12603	87511	187
188	99961	55979	83906	41570	88418	13371	31644	92267	92844	98253	188
189	31541	68322	33715	73510	01110	26036	99248	47424	38555	87958	189
190	34900	30367	23892	22642	82038	20552	35925	39647	90142	15046	190
191	94205	44990	91349	91531	64055	94886	55744	41662	72455	79836	191
192	84982	43754	09078	91460	99309	64563	61935	43214	99692	60101	192
193	04274	39885	97062	15927	12741	23963	02157	45642	02002	73102	193
194	34433	58119	70976	66159	17456	35073	85894	08935	27736	60578	194
195	97821	63898	17609	22580	42262	49780	75191	04821	87092	52186	195
196	20649	89577	45149	61825	60801	71346	67714	58156	00111	88857	196
197	64054	11178	87763	11005	09619	56917	38144	96278	47458	47826	197
198	50543	80232	69007	78996	35482	34818	43193	03514	72770	18560	198
199	84029	28230	93141	27645	21154	57594	42292	57147	46589	11512	199
200	68085	75447	30235	08190	42944	10980	35960	87461	79630	12952	200

TABLE **GG**   Power and sample size in anova

These charts provide estimates of power, $1 - \beta$, in an anova with numerator degrees of freedom, ν_1, from 1 to 8 and significance levels, α, of 0.05 and 0.01. They can also be used in an iterative fashion to determine the required sample size to achieve a desired power.

To estimate power, first turn to the graph corresponding to a particular value of ν_1 ($a - 1$ in a single-classification anova with a groups). Next, locate the curve corresponding to the denominator degrees of freedom, ν_2 [$a(n - 1)$ in a single-classification anova with equal sample size n], within the family of curves for the desired level of significance, α. Then compute ϕ, a standardized measure of the expected differences among means:

$$\phi = \sqrt{\frac{n\,\delta^2}{2a\,\sigma^2}}$$

where δ is the smallest expected difference between the two most dissimilar means and σ^2 is the within-groups variance (usually estimated by MS_{error}). The height of the curve at a point corresponding to an abscissa of ϕ gives the expected power. Note that there are separate scales for the curves corresponding to $\alpha = 0.05$ or 0.01. If one wishes to find the sample size needed to achieve a particular level of power, then one must try various values of n until one finds the smallest value that achieves the desired power.

For example, if $a = 6$, $n = 2$, $\alpha = 0.05$, $\delta = 10.0$, and $\sigma^2 = 5.46$, then $\nu_1 = a - 1 = 5$, $\nu_2 = a(n - 1) = 6$, and $\phi = 1.75$. The height of the curve for $\nu_2 = 6$ at $\phi = 1.75$ indicates a power value slightly greater than 0.55.

These graphs are used to determine power and estimate sample size in a single-classification anova (Box 9.14, Section 9.8).

These graphs were copied from charts in E. S. Pearson and H. O. Hartley (*Biometrika* **38**:112–130, 1951) with permission of the Biometrika Trustees.

TABLE GG Power and sample size in anova

Power = 1 - β

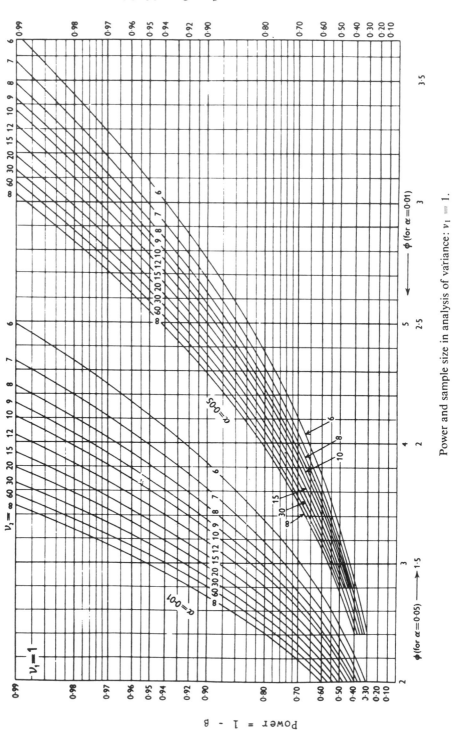

Power and sample size in analysis of variance: $\nu_1 = 1$.

TABLE GG Power and sample size in anova (*continued*)

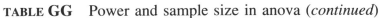

Power = 1 − β

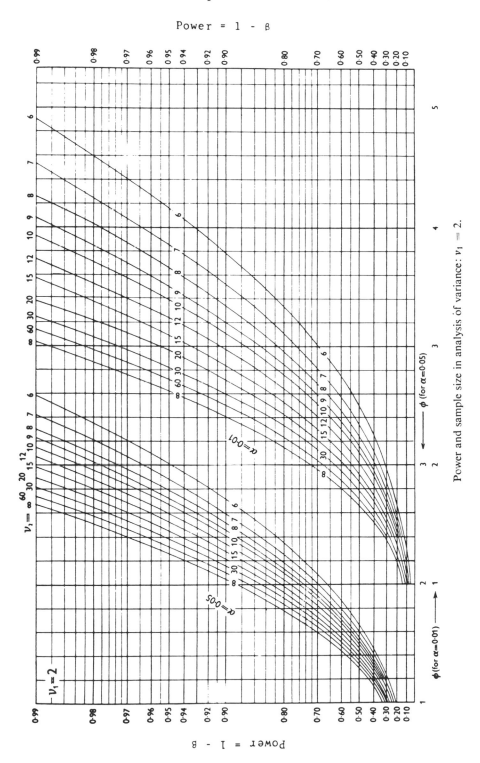

Power and sample size in analysis of variance: $v_1 = 2$.

190

TABLE GG Power and sample size in anova (*continued*)

Power = 1 - β

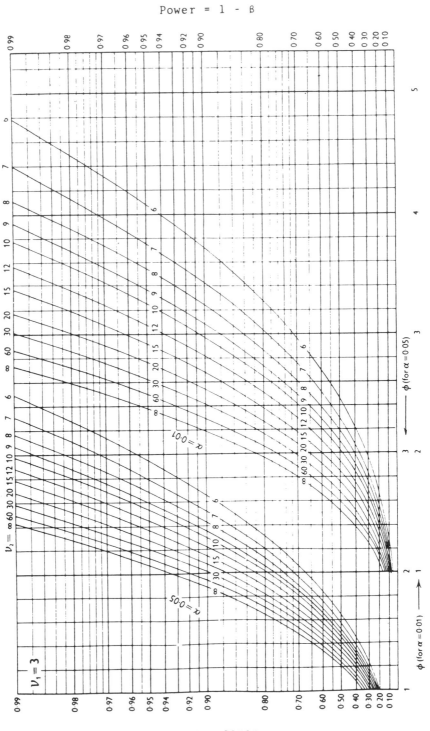

Power and sample size in analysis of variance: $\nu_1 = 3$.

TABLE GG Power and sample size in anova (*continued*)

Power = 1 - β

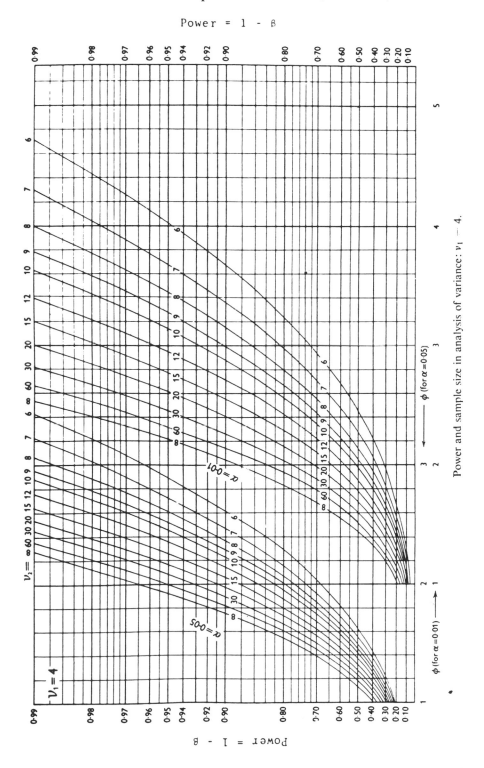

Power and sample size in analysis of variance: $\nu_1 - 4$.

Power = 1 - β

TABLE **GG**   Power and sample size in anova (*continued*)

Power = 1 - β

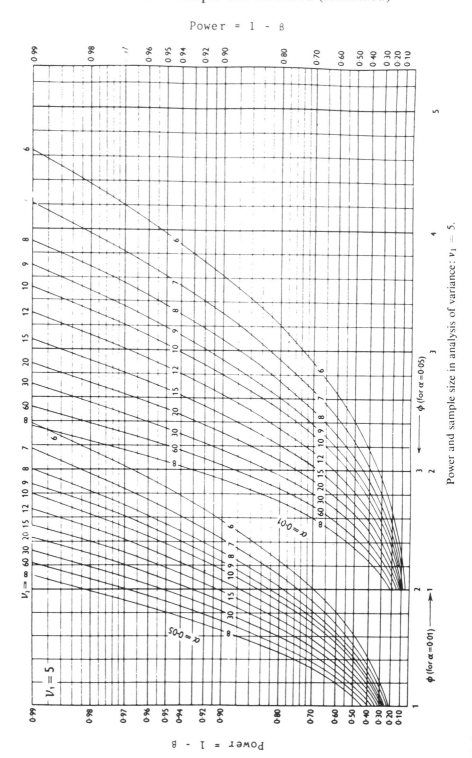

Power and sample size in analysis of variance: $\nu_1 = 5$.

TABLE GG   Power and sample size in anova (*continued*)

Power = 1 - β

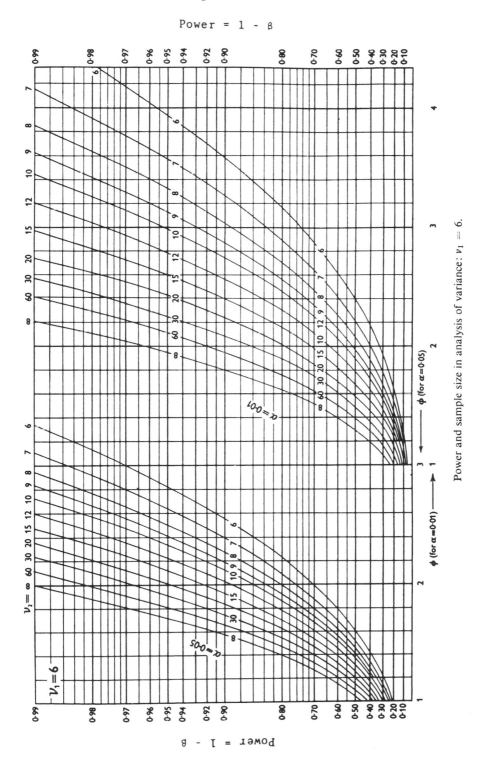

Power and sample size in analysis of variance: $\nu_1 = 6$.

TABLE **GG** Power and sample size in anova (*continued*)

Power = 1 - β

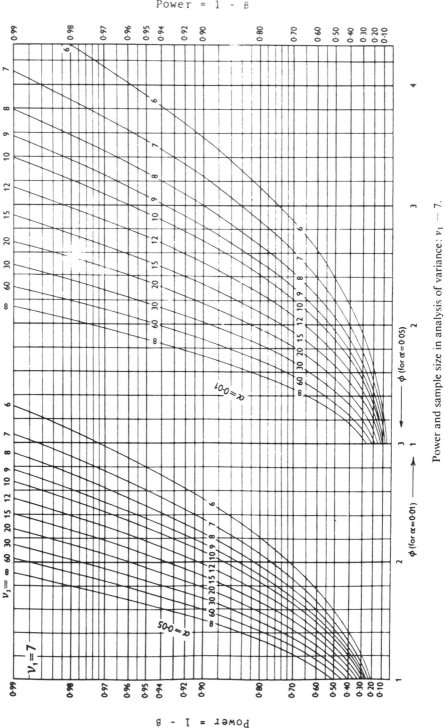

Power and sample size in analysis of variance: $v_1 - 7$.

Power = 1 - β

TABLE GG Power and sample size in anova (*continued*)

Power = 1 - β

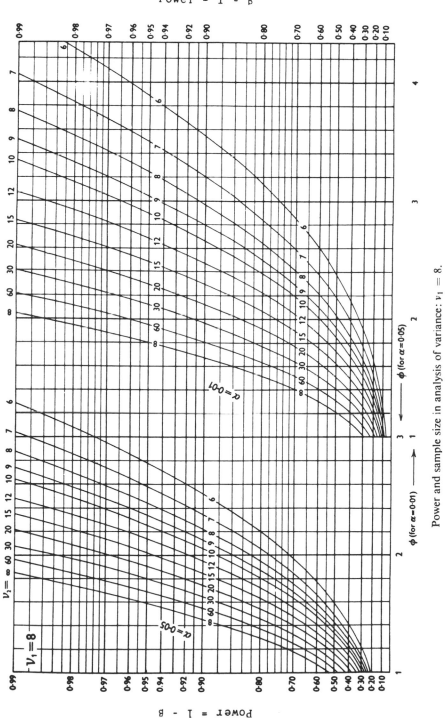

Power and sample size in analysis of variance: $\nu_1 = 8$.

TABLE **HH**   Critical values of $|1 - \eta/2|$, a test of serial independence

This table provides critical values of $C = |1 - \eta/2|$, a statistic to test for serial independence of a sequence of measurements for a continuous variable. The quantity η is the sum of the $n - 1$ squared differences between adjacent measurements in the ordered sample divided by the sum of squares. Critical values are provided for $\alpha = 0.1$, 0.05, 0.02, and 0.01 for sample sizes from $n = 8$ to $n = 50$. For larger samples one can take advantage of the fact that C is approximately normally distributed with a mean of 0 and a standard error, s_C, of $\sqrt{(n - 2)/(n - 1)(n + 1)}$. The appropriate factors are given in the last line of the table.

For a one-tailed test—e.g., if interested only in detecting deviations from independence in which adjacent measurements are more similar than one would expect as a result of chance (positive autocorrelation)—use $\alpha/2$ as the level of significance when C is positive.

This table is used to test the independence of a sequence of measurements (Box 13.1, Section 13.2).

The table was computed using a bisection method to determine values corresponding to the inverse of the incomplete beta function. Values for $n = 8$ to $n = 25$ and for $\alpha = 0.1$ and $\alpha = 0.02$ were compared with those of L. C. Young (*Ann. Math. Stat.* **12**:293–300, 1941). Small differences were found for $n = 8$ and $n = 9$ for $\alpha = 0.02$.

TABLE **HH** Critical values of $|1 - \eta/2|$, a test
of serial independence

	α			
n	0.1	0.05	0.02	0.01
8	.5088	.5874	.6681	.7207
9	.4878	.5650	.6454	.6954
10	.4689	.5444	.6241	.6729
11	.4517	.5257	.6043	.6529
12	.4362	.5085	.5860	.6343
13	.4221	.4928	.5691	.6170
14	.4092	.4784	.5534	.6008
15	.3973	.4651	.5389	.5857
16	.3864	.4528	.5254	.5717
17	.3764	.4414	.5128	.5585
18	.3670	.4308	.5010	.5461
19	.3583	.4209	.4900	.5346
20	.3502	.4117	.4797	.5236
21	.3426	.4030	.4700	.5133
22	.3355	.3948	.4608	.5036
23	.3288	.3871	.4521	.4944
24	.3224	.3799	.4439	.4857
25	.3164	.3730	.4361	.4774
26	.3108	.3665	.4287	.4695
27	.3054	.3603	.4217	.4619
28	.3003	.3544	.4150	.4548
29	.2955	.3488	.4086	.4479
30	.2908	.3434	.4025	.4413
31	.2864	.3383	.3966	.4350
32	.2822	.3333	.3910	.4290
33	.2781	.3287	.3856	.4232
34	.2742	.3241	.3805	.4177
35	.2705	.3198	.3755	.4123
36	.2670	.3157	.3707	.4072
37	.2635	.3117	.3661	.4022
38	.2602	.3078	.3617	.3974
39	.2570	.3041	.3574	.3928
40	.2539	.3005	.3533	.3883
41	.2510	.2970	.3493	.3840
42	.2481	.2937	.3455	.3798
43	.2454	.2905	.3417	.3758
44	.2427	.2874	.3381	.3719
45	.2400	.2843	.3346	.3680
46	.2376	.2814	.3312	.3644
47	.2351	.2785	.3279	.3608
48	.2328	.2758	.3247	.3573
49	.2305	.2731	.3216	.3539
50	.2283	.2705	.3186	.3507
>50	$1.6449s_C$	$1.9600s_C$	$2.3264s_C$	$2.5758s_C$

TABLE II C_n — Gurland and Tripathi's correction for the standard deviation

This table furnishes values of C_n, Gurland and Tripathi's correction for bias in estimates of the standard deviation σ. Values of C_n are given for all sample sizes n from 2 to 30.

For a sample size of $n = 10$, the correction factor C_n is 1.02811. When $n > 30$, a satisfactory approximation is $C_n \approx 1 + 1/4(n - 1)$.

The correction for bias is applied by multiplying s, the estimate of the standard deviation, by the correction factor C_n (as explained in Section 4.7).

The table was generated by computing the function

$$C_n = \frac{[(n - 1)/2]^{1/2}\Gamma[(n - 1)/2]}{\Gamma(n/2)}$$

in double precision arithmetic. The expression was developed by J. Gurland and R. C. Tripathi (*Am. Stat.* **25**:30–32, 1971).

n	C_n
2	1.25331
3	1.12838
4	1.08540
5	1.06385
6	1.05094
7	1.04235
8	1.03624
9	1.03166
10	1.02811
11	1.02527
12	1.02296
13	1.02103
14	1.01940
15	1.01800
16	1.01679
17	1.01574
18	1.01481
19	1.01398
20	1.01324
21	1.01257
22	1.01197
23	1.01142
24	1.01093
25	1.01047
26	1.01005
27	1.00966
28	1.00930
29	1.00897
30	1.00866

TABLE JJ   Some mathematical constants

The following mathematical constants are useful in a variety of statistical applications.

	Value	Reciprocal	$\log_{10}$
π	3.14159 26535 89793	0.31830 98861 83791	0.49714 98726 94134
2π	6.28318 53071 79586	0.15915 49430 91895	0.79817 98683 58115
$\pi/2$	1.57079 63267 94897	0.63661 97723 67581	0.19611 98770 30153
π^2	9.86960 44010 89359	0.10132 11836 42338	0.99429 97453 88268
$\sqrt{\pi}$	1.77245 38509 05516	0.56418 95835 47756	0.24857 49363 47067
$\sqrt{2\pi}$	2.50662 82746 31001	0.39894 22804 01433	0.39908 99341 79058
e	2.71828 18284 59045	0.36787 94411 71442	0.43429 44819 03252
e^2	7.38905 60989 30650	0.13533 52832 36613	0.86858 89638 06504
$\sqrt{e}$	1.64872 12707 00128	0.60653 06597 12633	0.21714 72409 51626
$\sqrt{2}$	1.41421 35623 73095	0.70710 67811 86548	0.15051 49978 31991
$\sqrt{3}$	1.73205 08075 68877	0.57735 02691 89626	0.23856 06273 59831
$\sqrt{10}$	3.16227 76601 68379	0.31622 77660 16838	0.50000 00000 00000
1 radian	57.29577 95130 82321°	0.01745 32925 19943	1.75812 26324 09172
ln 10	2.30258 50929 94046		